INFRARED OPTOELECTRONICS

OPTICAL ENGINEERING

Series Editor
Brian J. Thompson
Provost
University of Rochester
Rochester, New York

Laser Engineering Editor:
Peter K. Cheo
United Technologies Research Center
Hartford, Connecticut

Laser Advances Editor:
Leon J. Radziemski
New Mexico State University
Las Cruces, New Mexico

Optical Materials Editor:
Solomon Musikant
Paoli, Pennsylvania

1. Electron and Ion Microscopy and Microanalysis: Principles and Applications, *by Lawrence E. Murr*
2. Acousto-Optic Signal Processing: Theory and Implementation, *edited by Norman J. Berg and John N. Lee*
3. Electro-Optic and Acousto-Optic Scanning and Deflection, *by Milton Gottlieb, Clive L. M. Ireland, and John Martin Ley*
4. Single-Mode Fiber Optics: Principles and Applications, *by Luc B. Jeunhomme*
5. Pulse Code Formats for Fiber Optical Data Communication: Basic Principles and Applications, *by David J. Morris*
6. Optical Materials: An Introduction to Selection and Application, *by Solomon Musikant*
7. Infrared Methods for Gaseous Measurements: Theory and Practice, *edited by Joda Wormhoudt*
8. Laser Beam Scanning: Opto-Mechanical Devices, Systems, and Data Storage Optics, *edited by Gerald F. Marshall*
9. Opto-Mechanical Systems Design, *by Paul R. Yoder, Jr.*

10. Optical Fiber Splices and Connectors: Theory and Methods, *by Calvin M. Miller with Stephen C. Mettler and Ian A. White*
11. Laser Spectroscopy and Its Applications, *edited by Leon J. Radziemski, Richard W. Solarz, and Jeffrey A. Paisner*
12. Infrared Optoelectronics: Devices and Applications, *by William Nunley and J. Scott Bechtel*
13. Integrated Optical Circuits and Components: Design and Applications, *edited by Lynn D. Hutcheson*

LASER HANDBOOKS—*Edited by Peter K. Cheo*

1. Handbook on Molecular Lasers

Other Volumes in Preparation

INFRARED OPTOELECTRONICS

Devices and Applications

WILLIAM NUNLEY
J. SCOTT BECHTEL
TRW Electronic Components Group
Carrollton, Texas

MARCEL DEKKER, INC.　　　　　　　　New York and Basel

Library of Congress Cataloging-in-Publication Data

Nunley, William, [date]
   Infrared optoelectronics.

   (Optical engineering; 12)
   Bibliography: p.
   Includes index.
   1. Optoelectronic devices. 2. Infrared technology.
I. Bechtel, J. Scott, [date]. II. Title.
III. Series: Optical engineering (Marcel Dekker, Inc.);
12.
TA1750.N86 1987   621.36'2   86-24217
ISBN 0-8247-7586-4

COPYRIGHT © 1987 by WILLIAM NUNLEY AND J. SCOTT BECHTEL
ALL RIGHTS RESERVED

Neither this book nor any part may be reproduced or transmitted in any form or by any means, electronic or mechanical, including photocopying, microfilming, and recording, or by any information storage and retrieval system, without permission in writing from the authors and the publisher.

MARCEL DEKKER, INC.
270 Madison Avenue, New York, New York 10016

Current printing (last digit):
10 9 8 7 6 5 4 3 2 1

PRINTED IN THE UNITED STATES OF AMERICA

About the Series

Optical science, engineering, and technology have grown rapidly in the last decade so that today optical engineering has emerged as an important discipline in its own right. This series is devoted to discussing topics in optical engineering at a level that will be useful to those working in the field or attempting to design systems that are based on optical techniques or that have significant optical subsystems. The philosophy is not to provide detailed monographs on narrow subject areas but to deal with the material at a level that makes it immediately useful to the practicing scientist and engineer. These are not research monographs, although we expect that workers in optical research will find them extremely valuable.

Volumes in this series cover those topics that have been a part of the rapid expansion of optical engineering. The developments that have led to this expansion include the laser and its many commercial and industrial applications, the new optical materials, gradient index optics, electro- and acousto-optics, fiber optics and communications, optical computing and pattern recognition, optical data reading, recording and storage, biomedical instrumentation, industrial robotics, integrated optics, infrared and ultraviolet systems, etc. Since the optical industry is currently one of the major growth industries this list will surely become even more extensive.

Brian J. Thompson
Provost
University of Rochester
Rochester, New York

Preface

Modern day optoelectronics utilizes an infrared energy emitting diode and a silicon photosensor to create an optical pathway. An opaque object interrupting this energy path creates a switch action. This switch is different from mechanical switches in that there are no mechanical contacts to make or break. Optical switches are widely used in computer peripheral and office equipment machines as well as other industrial applications for motion detectors or travel limit switches. In 1986, sales of optical transmissive and reflective switches, optocouplers, and discrete infrared emitting diodes and photosensors constituted a $400 million market. This market is growing at more than 10% a year, and new applications are being developed at a phenomenal rate.

Several companies have published data books, application notes, and handbooks that served as general background for this book. This book is intended as a primary reference and introduction to infrared optoelectronics. The reader should be able to develop a basic understanding of the component and how and where to use it through study of this professional reference.

Part I deals with the transmitter side of the switch, which is a solution-grown epitaxial diode or IRED. The two most common materials used to manufacture these transmitters or infrared emitters are gallium arsenide (GaAs) and gallium aluminum arsenide (GaAlAs). They are discussed in terms of the basic semiconductor, the package, characterization, and reliability. Brief reference is made to other types of IREDS.

Part II deals with the receiver or photosensor. This is usually made from silicon owing to the close spectral matching between silicon and gallium arsenide or gallium aluminum arsenide and the relative ease of fabricating the photosensor using known and widely dispersed technology. Several types of photodiodes through the phototransistor and the optical integrated circuit are presented. Advantages and disadvantages of each are discussed. There is also a brief discussion of special function photo integrated circuits (photo ICs).

Part III deals with the coupled emitter/photosensor pair with emphasis on digital application in both the transmissive and reflective modes. Part IV deals with the optical isolator, or DC transformer.

Part V deals with open air communications, fiber optic communication, and long-range beam interrupt or surveillance type applications. Since this primarily concerns pulse type transmission and reception, pulse techniques for the IRED and receiver are also covered in this section. Part VI deals with a number of different applications for different market segments, such as automotive, medical electronics, telecommunications, consumer, and computer peripheral equipment. Drive circuits for the IRED transmitters and interface circuitry for the photosensor receivers are also discussed. Part VII is a glossary of the symbols and terms unique to optoelectronics.

Practicing engineers as well as undergraduate or graduate students should find this book a practical introduction to optoelectronics. The discussion is based on many years of answering the questions most commonly posed by engineers developing designs using these products. Therefore a practical rather than a theoretical approach is emphasized. As such, it offers a convenient engineering reference.

The authors wish to thank Charolette Smith for her patience and understanding in the preparation of the typed manuscript as well as the incorporation of the many changes. A special note of thanks goes to Billy G. Davis for his creative and diligent work on the illustrations. The authors also wish to express their gratitude to the technical staff of TRW Optoelectronics Division for both their proofreading time and the technical bulletins and characterization data that comprise the bulk material of this reference book. It is our sincere wish that the information presented will provide the engineer with answers to most of the commonly asked questions on optoelectronics.

William Nunley
J. Scott Bechtel

Contents

About the Series   iii

Preface   v

Part I   The Source (Infrared Emitting Diodes)

1. BASIC THEORY   3
   - 1.1 Introduction   3
   - 1.2 P-N junction injection electroluminescence   3
   - 1.3 Relative efficiency   5

2. IRED FABRICATION   7
   - 2.1 Introduction   7
   - 2.2 Solution grown epitaxial GaAs   7
   - 2.3 Solution grown epitaxial GaAlAs   10

3. IRED PACKAGING   14
   - 3.1 Introduction   14
   - 3.2 Techniques for improving photon emission efficiency   14
   - 3.3 Packaging the IRED   19
   - 3.4 Characterization of the packaged IRED   26
   - 3.5 Understanding thermal impedance   45
   - 3.6 Understanding the measurement of radiant energy   53
   - 3.7 Reliability   62

Part II   The Receiver (Silicon Photosensor)

4. THE PHOTODIODE   69
   - 4.1 Basic theory   69
   - 4.2 Characterization   72

	5.	THE PHOTOTRANSISTOR AND PHOTODARLINGTON	81
		5.1 Basic theory	81
		5.2 Characterization	85
	6.	THE PHOTOINTEGRATED CIRCUIT	99
		6.1 Basic theory	99
		6.2 Characterization	109
	7.	SPECIAL FUNCTION PHOTO ICs	113
		7.1 Basic theory	113
		7.2 Characterization	114
		7.3 Triac driver photosensors	121
PART III	The Coupled Emitter (IRED) Photosensor Pair		
	8.	THE TRANSMISSIVE OPTICAL SWITCH	127
		8.1 Electrical considerations	127
		8.2 Mechanical considerations	149
	9.	THE REFLECTIVE OPTICAL SWITCH	154
		9.1 Electrical considerations	154
		9.2 Mechanical considerations	162
PART IV	The Optical Isolator or Coupler		
	10.	ELECTRICAL CONSIDERATIONS	167
		10.1 Background	167
		10.2 Function	169
		10.3 Different types	175
	11.	MECHANICAL CONSIDERATIONS	184
		11.1 Fabrication techniques	184
		11.2 Other mechanical considerations	188
PART V	Open Air and Fiber Optic Communication		
	12.	FIBER OPTIC COMMUNICATION	193
		12.1 Basic theory	193
		12.2 Background	194
		12.3 Fiber optic technical barriers	195
	13.	OPEN AIR COMMUNICATION	197
		13.1 Basic theory	197
		13.2 Background	199

Contents ix

14. PULSE OPERATION 202
    14.1 The IRED 202
    14.2 The photosensor 207

PART VI Optoelectronics Applications

15. DRIVING THE IRED 213
    15.1 Introduction 213
    15.2 Variables 213
    15.3 Linear operation 215

16. INTERFACING TO THE PHOTOSENSOR 217
    16.1 The photodiode 217
    16.2 The phototransistor and photodarlington 217
    16.3 The photointegrated circuit 223

17. COMPUTER PERIPHERAL AND BUSINESS EQUIPMENT APPLICATIONS 224
    17.1 Winchester drive 224
    17.2 Floppy disks 224
    17.3 Printer 224
    17.4 Keyboard and touchscreen 226
    17.5 Computer mouse 228
    17.6 Card reader/tape reader 229
    17.7 Tape drive 229
    17.8 Photocopier 230
    17.9 Miscellaneous optical sensors 231
    17.10 Optical couplers 232

18. INDUSTRIAL APPLICATIONS 233
    18.1 Safety-related optical sensors 233
    18.2 Security and surveillance systems 233
    18.3 Mechanical aids, robotics 233
    18.4 Miscellaneous optical sensors 240
    18.5 Optical couplers 240
    18.6 Contour sensors 240

19. AUTOMOTIVE APPLICATIONS 242
    19.1 Existing applications 242
    19.2 Future applications 242

20. MILITARY APPLICATIONS 246
    20.1 Military applications for optical sensors 246

21.	CONSUMER APPLICATIONS		249
	21.1	TV games or toy controls	249
	21.2	TV and CATV remote communications	249
	21.3	Video disk or record control	252
	21.4	Dollar bill changer	252
	21.5	Smoke detectors	252
	21.6	Coin changers and slot machines	253
	21.7	Camera applications	253
	21.8	Optical golf game	254
	21.9	Pool cue ball sensor	254
	21.10	Household appliance cycle control	254
	21.11	Three-dimensional video or film	256
22.	MEDICAL APPLICATIONS		257
	22.1	Introduction	257
	22.2	Intravenous drop monitor	257
	22.3	Pulse rate detection	257
	22.4	Measurement of glaucoma	257
	22.5	Optical switch for paraplegics	259
23.	TELECOMMUNICATIONS		260
			260

Glossary	261
Bibliography	269
Index	271

INFRARED OPTOELECTRONICS

I
THE SOURCE (INFRARED EMITTING DIODES)

1
Basic Theory

1.1 INTRODUCTION

This chapter is devoted to a basic understanding of the transmitting portion of the optical switch. This infrared source typically takes the form of an infrared light-emitting diode. Other terms sometimes used by engineers to refer to these devices include the following: emitters, IR emitters, LED, IRED, or IRLED. While all terms refer to essentially the same device, this book uses infrared emitting diode and IRED because it is by far the most commonly used terminology. The reader will receive an understanding of how the energy is transformed from current to infrared energy, why the materials for normally used generation of this energy were selected, how the infrared emitting diode is made, how the emitting wavelength is controlled, and how the energy is emitted. The discussion will not delve deeply into semiconductor physics, because only a conceptual understanding of the element is required for successful use of the device in applications, which is the major goal of this reference book.

1.2 P-N JUNCTION INJECTION ELECTROLUMINESCENCE

A P-N junction can be formed in a semiconductor material by doping one region with donor atoms and an adjacent region with acceptor atoms. This produces a nonuniform distribution of "impurities" with an abrupt change from one type of doped material to another. When the donor atoms dominate, the material is known as N-type and there is an excess of N-type atoms or majority carriers. For N-type material, the majority carriers are negatively charged and are electrons. When the acceptor atoms dominate, the material is known as P-type and the majority carriers are positively charged and are holes. Figure 1.1 shows pictorially the P-N junction with the excess of negatively charged electrons and the positively charged holes.

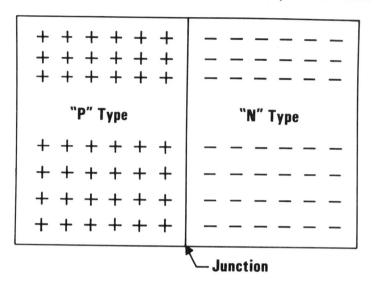

Figure 1.1 The P-N junction may be described as an abrupt change from material containing a majority of positively charged carriers or holes to material containing an excess of negatively charged carriers or electrons.

When the junction is formed these electrons and holes will flow in opposite directions across the junction (without applied bias) until equilibrium is reached. This gives rise to a built-in potential barrier. If an external electrical bias is applied across the junction that counteracts this built-in potential, additional electrons and holes will be injected or flow across the P-N junction. These carriers will then recombine by either a *radiative* or *nonradiative* process. In the case of a radiative process, the recombination requires that the energy level of the electron drop in order to facilitate recombination with a hole. The electron sheds this excess energy in the form of a discrete amount of infrared energy known as a photon. In a nonradiative process, excess energy is still released but takes the form of heat and is quantified as *phonons*. In both cases, this recombination occurs in close proximity (a few diffusion lengths) to the P-N junction.

The maximum possible energy of the emitted photons is determined by the band-gap energy of the solid in which the P-N junction is formed. There are numerous elements and elemental compounds that have band-gap energies that lie in the region from ultraviolet to infrared. However, very few of these materials are viable candidates for practical LEDs. Table 1.1 lists several materials utilized for commercial devices.

1. Basic Theory

Table 1.1 Some of the Materials Utilized for LED Devices

Material	Band gap energy	Emission wavelength
	Electron volts (eV)	Nanometers (nm)
Gallium arsenide (GaAs)	1.33	930
Gallium aluminum arsenide (GaAlAs)	1.41	880
Gallium arsenide phosphide (GaAsP)	1.91	650
Gallium phosphide (GaP)	2.21	560

The band gap energy in Table 1.1 is calculated from the formula:

$$E = \frac{1240}{\lambda} \quad \text{electron volts}$$

where λ is the emission wavelength and E is the energy transition in electron volts

EXAMPLE: GaAs; E = 1240/930 = 1.33 eV

1.3 RELATIVE EFFICIENCY

The percentage of the current that results in recombinations that give rise to photons of the desired wavelength is a measure of the internal conversion efficiency of the P-N diode. A material with low internal conversion efficiency would offer little practical interest as an electroluminescent device. However, even a material with a high internal conversion efficiency may not be useful if the emitted photons cannot be efficiently "emitted" from the diode structure or "coupled" to the external environment. Two major factors control the internal to external coupling coefficient. One factor is the opacity of the diode material. This is directly related to the reabsorption of emitted photons. The second factor is the internal reflection at the interface of the semiconductor crystal and the encapsulation material. This may cause the photon to be reflected back into the crystal and subsequently reabsorbed. Table 1.2 shows the relative efficiency of some LEDs as a percentage of "gap" efficiency.

Table 1.2 Relative Efficiency of Some LEDs

Material	Relative output efficiency (% gap efficiency)	Wavelength emission (nm)
GaP	100	560
GaAs	400	930
GaAlAs	800	880

This efficiency increases if there is a decrease in the crystal's probability of reabsorbing or reflecting back photons emitted at the junction. If more photons escape from the diode material, the relative output efficiency increases.

It becomes very obvious that the GaAlAs and GaAs IREDs are preferred transmitters because of the efficiency with which they emit infrared energy. However, to be used commercially, it is necessary to have detectors available that respond to the same wavelength of infrared. In other words, the spectral response of the detector must form a reasonable match to the emission wavelength of the material chosen for the IRED. Silicon is the material normally utilized since its peak absorption falls in the 750 to 950 nm range. This will be discussed in more detail in Part II, which deals with the photosensor.

The selection of GaAs and GaAlAs for use in optical switches is primarily because of improved efficiency (conversion of input power to usable output energy) over GaAsP and GaP; however, its choice is also contingent upon the availability of photosensors capable of detecting the wavelengths emitted.

2
IRED Fabrication

2.1 INTRODUCTION

In Section 1.2 the P-N junction was described as semiconductor material with adjacent regions where one region was doped with donor atoms and one was doped with acceptor atoms. Figure 2.1 shows the cross-sectional profile of a diffused junction.

In Region 1 the N-type impurities (donor atoms) dominate while in Region 2 the P-type impurities (acceptor atoms) dominate. The concentration at any given point is the sum of the negative N-type and the positive P-type atoms. The junction is the point at which the negative N-type atoms equal the positive P-type atoms so the net sum is zero. The shape of the impurity concentration profile will change as a function of how the impurities are introduced and the treatment of the material after the doping or the addition of the impurities takes place. The P-N junction will be the point at which the net concentration is zero.

It is important to note that there are N-type and P-type impurities on both sides of the junction. In the region where the N-type impurities dominate the N-type atoms are known as majority carriers and the P-type atoms are known as minority carriers. The reverse is true on the P-type side of the junction.

2.2 SOLUTION GROWN EPITAXIAL GaAs

The epitaxial technique adds more semiconductor material to the material originally present. This growth or addition of semiconductor material having the same crystalline structure as the original material allows flexibility in both how the donor or acceptor atoms are introduced and what their concentration profile looks like. The particular epitaxial technique described will be *solution grown*, which means the growth occurs from a solution source located adjacent to the semiconductor material. The single crystal GaAs slice, which has majority donor or N-type impurity atoms, is placed in a suitable carrier and raised to a temperature in excess of 900° C. The melt (solution containing gallium, N, and P-type impurities, and polycrystalline GaAs)

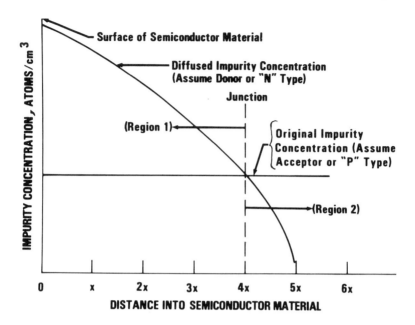

Figure 2.1 Cross-sectional profile of diffused junction. The starting material contains a uniform concentration of P-type impurities to which N-type impurities have been added by the diffusion process. The concentration is highest near the surface and drops to the same level as the P-type doping below the surface. It is here that the P-N junction is formed. Beyond the junction depth, the original P-type impurities remain as the majority carriers.

is then placed on the slice and the epitaxial layer of GaAs starts to grow. The melt has both donor (N-type) and acceptor (P-type) impurity atoms present. The temperature is gradually reduced. While growth continues, the donor (N-type) atoms that dominate will decrease in concentration. The acceptor (P-type) atoms correspondingly increase in concentration. At approximately 900° C the donor (N-type) atoms and the acceptor (P-type) atoms become equal in concentration and the junction is formed.

The temperature continues to decrease and the concentration of the acceptor (P-type) atoms increase. Figure 2.2 shows these changes graphically. The temperature is decreased between points (1) and (3) where growth starts and stops.

2. IRED Fabrication

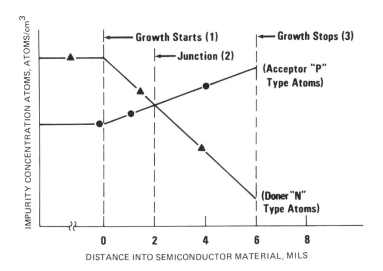

Figure 2.2 Impurity profile GaAs IRED. The concentration of N-type impurities slowly diminishes as the crystal layer begins to grow. Simultaneously, the concentration of P-type impurities increases. At about 900° C, the concentrations become equal and the P-N junction is formed. Below 900° C, the crystalline layer grows with a majority of P-type impurities.

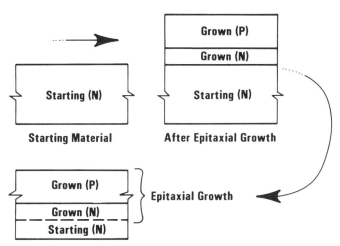

Figure 2.3 GaAs slice with solution grown GaAs epitaxial layer. N-type GaAs is used as the substrate to grow additional N-type material transitioning to P-type material to form the P-N junction. Excess N-type material is then removed prior to metallization used to form electrical contact pads.

The starting slice, which was approximately 14 mils thick, has been increased in thickness to approximately 20 mils by the epitaxial growth. The junction is approximately 4 mils down from the top. Figure 2.3 shows the profile of these layers including the profile after removal of the excess starting material. Approximately 12 mils of the original starting slice is removed.

The final processing step is to add metal contacts to the P-type side and N-type side of the slice, prior to separation into individual IREDs.

2.3 SOLUTION GROWN EPITAXIAL GaAlAs

GaAlAs IREDs are made in a similar manner. The process requires more stringent controls because there is an additional variable of the aluminum added to the melt or solution source. The growth process is essentially the same with minor changes to the starting temperature. As the temperature is decreased the epitaxial layer first grows N-type GaAlAs. The aluminum that forms in the crystalline structure has the heaviest concentration at the initial starting crystal/growth interface and decreases along a logarithmic curve. The doping level is controlled to be approximately 5% by volume at the P-N junction.

The doping level of the aluminum atoms at the junction controls the wavelength emitted. Increasing it will shorten the wavelength and decreasing it will lengthen the wavelength (variation in emitted wavelength can change from approximately 800 nm to 930 nm). The 5% aluminum concentration by volume gives emission at 875 nm. This provides a close spectral match to the silicon phototransistor discussed in Part II. Figure 2.4 shows the same type of concentration profile as was shown in the section on GaAs (2.1) but with the addition of the aluminum. The temperature is decreased between points (1) and (3) where growth starts and stops.

The starting slice, which was approximately 14 mils thick, has been increased in thickness to approximately 22 mils by the epitaxial growth. The junction is approximately four mils down from the top. Figure 2.5 shows the profile of these layers, including the profile after removal of the excess starting material (approximately 14 mils, or the original starting slice, is removed). The addition of the aluminum to the grown N-type region allows the use of an etchant for excess material removal. This chemically attacks the GaAs N-type material but not the material containing aluminum.

The three major differences between GaAs and GaAlAs are shown in Table 2.1.

The epitaxial growth stage for GaAlAs diodes requires more stringent controls of the total growth phase. This mandates increased processing steps and more sophistication in the process and equipment. The excess material removal of the starting GaAs slice is less diffi-

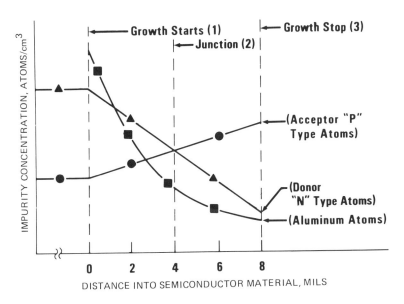

Figure 2.4 Impurity profile of GaAlAs IRED. Growth of GaAlAs diode material differs from GaAs epitaxial growth due to the use of aluminum in the solution source. Control of the aluminum concentration at the P-N junction is critical in determining wavelength.

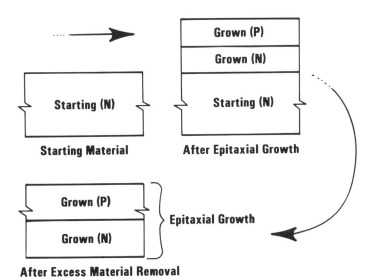

Figure 2.5 GaAs slice with solution grown GaAlAs epitaxial layer. GaAlAs contains a similar impurity profile to GaAs diodes, except for the addition of aluminum to the growth layers. This addition allows for removal of the excess N-type starting material through the use of lower cost etching techniques. Device efficiency is also improved.

12   The Source (Infrared Emitting Diodes)

Table 2.1   Major Differences Between GaAs and GaAlAs

	Epitaxial growth	Excess mtl removal	Doping
GaAs	Relatively simple	More complex	Relatively simple
GaAlAs	More complex	Relatively simple	More complex

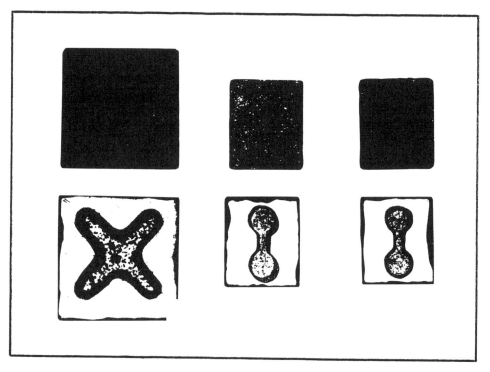

Figure 2.6   Assorted GaAs and GaAlAs chips. The chip size is directly related to its maximum current rating. The contact pad design affects current and forward voltage ratings.

2. IRED Fabrication

cult with GaAlAs, because selective etching can be used to remove GaAs at a far faster rate than GaAlAs. Mechanical abrasion, which is a more expensive process, is required to remove excess material on the GaAs wafers.

After removal of the excess material, metallization is applied to the complete P-type side. This side will subseqently become the mounting surface of the chip. The N-type side then becomes the top side and has only a localized contact added. This provides the necessary electrical contact but minimizes the area covered in order to allow the maximum amount of infrared energy to escape. Some manufacturers use the reverse process where the N-type side is mounted down rather than the P-type side. Figure 2.6 is a photograph of the top and bottom surfaces of a variety of GaAs and GaAlAs chips. The small chips are 0.012 in. long × 0.010 in. wide × 0.008 in. high while the larger chips are 0.016 in. long × 0.016 in. wide × 0.008 in. high.

3
IRED Packaging

3.1 INTRODUCTION

In Section 1.2 the mechanism of photon formation was described, and in Secton 1.3 the emission of these photons was described. A brief review will clarify this mechanism. Figure 3.1 shows the bias arrangement for forward conducton of a P-N junction.

Figure 3.2 optically models this recombination as a point source. It should be noted that the planes formed across the cross section parallel to the junction and within a few diffusion lengths of the junction will contain a very large number of these point sources emitting energy. By reviewing a single point source the overall mechanism can be understood.

Figure 3.3 shows how these get to the outside world.

3.2 TECHNIQUES FOR IMPROVING PHOTON EMISSION EFFICIENCY

There are many techniques that have been practiced for improving the percentage of photons that escape from the semiconductor material. Figure 3.4 shows an early technique used by an IRED manufacturer.

Two IREDs are fastened together with the P-type sides common. They are then tumbled in a process similar to polishing gems. When a sphere is formed, the process is stopped and the unit is segregated into two half-spheres. The two units that result will have improved emission efficiency since the number of photons approaching the N_1/N_2 interface at less than the critical angle has been increased.

Another similar technique is practiced by most manufacturers of IREDs. It is simply an artificial roughening of the surface that creates more surface area. This, in turn, allows more photons to escape. This is shown in Figure 3.5.

A common technique is to mount the IRED in a well or cavity contoured in such a fashion that the photon energy escaping the sides of the chip will be reflected toward the receiving area. A silicone compound (Jello-like consistency) can be placed over the IRED chip. By virtue of refractive index matching, this increases the size of the

3. IRED Packaging

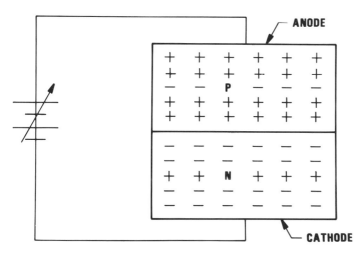

Figure 3.1 Bias arrangement for forward conduction of a P-N junction. When the supply or battery voltage is below approximately 0.9 V, conduction does not occur. Once conduction starts, recombination will occur. The most efficient recombination occurs within a few diffusion lengths of the P-N junction in both the P-type and the N-type regions. The energy release in this recombination is a radiative process generating photons.

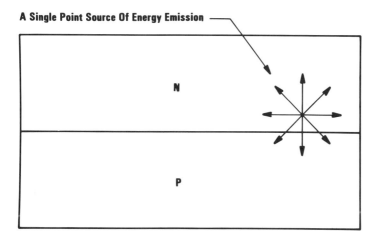

Figure 3.2 Energy emission from a point source near the P-N junction. The photons created during this recombination process will radiate out in all directions from the point source.

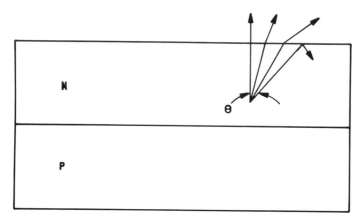

Figure 3.3 Energy emission from the IRED to the outside world. Some of the photons pass vertically through the N_1/N_2 (IRED/outside world) interface. N_1 is the normal method of referring to material the energy is moving "from" or "in" with N_2 being the adjacent material. As the angle between the vertical and the horizontal path of the photons is increased, the photon path becomes more and more bent at this N_1/N_2 interface. At a point called the *critical angle*, the photons do not escape and are reflected back into the semiconductor material. The critical angle with GaAs or GaAlAs as the N_1 material and air as the N_2 material approximates 16°. This critical angle will change as the N_1 and N_2 materials change. The photons that escape become potentially useful energy for transfer while the remaining photons are reabsorbed and lost within the IRED.

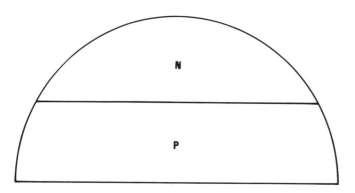

Figure 3.4 Contouring the N_1/N_2 interface to improve efficiency. The problem created by critical angle reflections is lessened by altering the device geometry. A hemispherical shape increases the number of escaping photons by lowering the probability of photons encountering the interface at an angle in excess of the critical angle.

3. IRED Packaging

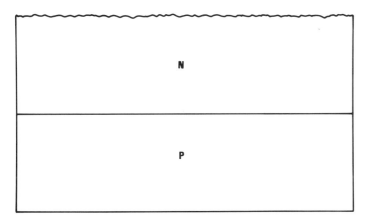

Figure 3.5 Increasing surface area by roughening the chip.
A second version of efficiency improvement by altering chip geometry is to roughen the surface of the chip. From a cost standpoint, this is a more practical method of increasing efficiency than that shown in Figure 3.4.

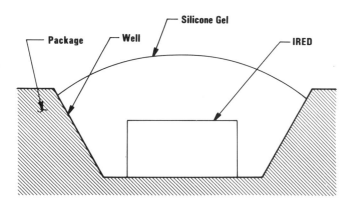

Figure 3.6 IRED mounted in a well with silicone gel coverage. Usable energy emitted from the sides of the IRED is directed upward by the reflective well, where it encounters the hemispherical interface of silicone gel and air (or plastic encapsulation material). The semiconductor/silicone gel interface has a larger critical angle, resulting in 30% greater output.

critical angle to approximately 22°. This results in approximately 30% more photons being able to escape and become potential usable energy. Figure 3.6 shows the IRED mounted in a well with the silicone gel compound over it.

As mentioned in Section 3.1, the maximum generation of photons occurs adjacent to the P-N junction. If we assume an IRED chip of 0.016 in. long × 0.016 in. wide × 0.008 in. high with the junction centered in the chip, the top surface has a radiating area of 256 sq. mils, the regions on the sides have a radiating area of 0.008 in. × 0.016 in. × 4, or 512 sq. mils. The longer the distance the photons have to travel toward the sides to escape means a larger number will be reabsorbed (the vertical distance is 0.004 in., while the horizontal distance is an average of 0.008 in. to 0.010 in.). The net result of this geometry is that the top radiation will be approximately 60% of the total with the side radiation being 40% of the total. These ratios are significantly altered as the geometry of the chip changes. The addition of the well will improve the effective energy radiated by approximately 40% by utilizing the energy radiated from the sides of the IRED.

A significant difference in effective energy radiated also occurs as different materials are used for the package. Figure 3.7 shows the package outline for both a hermetic metal and a plastic encapsulated unit.

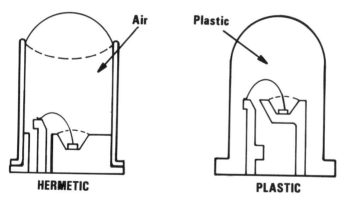

Figure 3.7 Package outline for hermetic and plastic case. The hermetic package has four interfaces that the infrared energy must traverse in order to escape the package (the chip/silicone, the silicone/air, the air/glass, the glass/air). In the plastic package there are only three interfaces (chip/silicone, silicone/plastic, and plastic/air). In actual practice, the silicone/plastic interface has little effect on transmission because the refractive indices for the silicone and the plastic are virtually the same. This reduction by two interfaces in the plastic package allows an improvement of 40 to 100% in escaping IR energy.

3. IRED Packaging

3.3 PACKAGING THE IRED

Packaging of the IRED consists of four basic operations. The chip is placed in the well and the mounting provides both an electrical path for current flow as well as a thermal path to remove the heat from the chip. There are three different techniques commonly used for this mounting.

Conductive epoxy mount: A paste containing metallic particles such as silver or gold contained in a carrier or binder (with a thinner that allows the material to be dispensed) is placed in the well. The chip is then mounted into this paste. Heat causes the paste to cure, driving off the thinner and forming a mechanical bond to both the chip and the mount area.

Alloy mount: Layers of deposited metal on the P-type side of the chip allow ohmic and mechanical contact to the chip as well as ohmic and mechanical contact to the mount area. This is different from the conductive epoxy mount in that the metal deposited on the P-type region of the chip actually penetrates into the chip and forms an alloyed region.

Solder mount: This is similar to conductive epoxy mount in that the metallization on the P-type region of the chip and the metallization on the mount area is mechanically and electrically bonded by

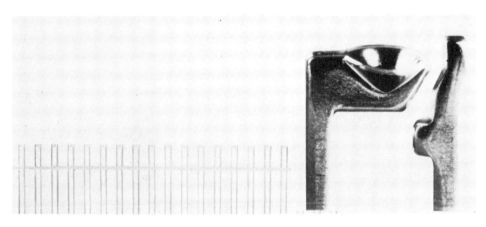

Figure 3.8 IRED 20 unit lead frame with detail enlargement of IRED chip mounting area. Production economies are gained by processing twenty IREDs per lead frame. All three mounting methods are possible with this configuration.

the solder flow. The solder forms a contact to the metallizaton and does not penetrate into the chip itself.

Each of these techniques has advantages and disadvantages. If proper controls are used, the manufacturer can select the optimal system for volume/cost tradeoffs. Figure 3.8 shows an IRED lead frame of 20 units and a detailed enlargement of the chip mounting area.

The pictures shown are for a plastic packaged IRED with conductive epoxy mount. The process would be the same in principle for any of the three methods of mounting discussed previously.

The mounting may be done with any one of three levels of mechanical assistance. The operation and end product will be essentially the same but the output per hour will be significantly different. Figure 3.9 shows three different mount machines with varying output rates. The differences in operating speed are the result of mechanical aids for the operator. Note the output rate increases with the increase in mechanical aids.

The production rates given in Figure 3.9 show a dramatic difference in throughput. If we were to assume a hypothetical labor rate of $4.00/hour and an overhead rate of 200% (cost of benefits, depreciation, expendable supplies, space, supervision, etc.), the burdened cost of labor would be $12.00/hour. Utilizing this number with the rates shown in Figure 3.9 we see:

100 units/hr = $\frac{\$12.00}{100}$ or $0.12 /unit (Figure 3.9a)

500 units/hr = $\frac{\$12.00}{500}$ or $0.024/unit (Figure 3.9b)

5000 units/hr = $\frac{\$12.00}{5000}$ or $0.0024/unit (Figure 3.9c)

This is not a linear function, because the overhead rates will change as the equipment gets faster and more expensive (the depreciation will increase while the labor cost and associated cost of benefits will decrease). This usually means higher overhead rates due to increased amortization costs per unit. The net result in overall cost, nevertheless, favors very heavily the higher units per hour per operator.

The second operation is the bonding that allows contact to be made to both the N-type region of the IRED and the cathode lead exiting the package. There are three levels of automation possible for this operation, very similar to rates to the previously discussed mount operation. Mounting and bonding are potentially the highest labor-consuming operations in the entire process of fabrication of the IRED.

There are two different bonding techniques used to make electrical and mechanical contact to the chip and lead frame: thermocompression and ultrasonic bonding. Thermocompression bonding is normally

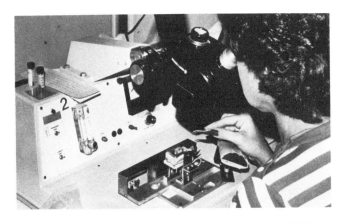
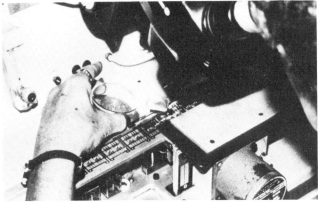

Figure 3.9   Three levels of mechanical mounting aid. A variety
of semiconductor manufacturing equipment is available for use
fabricating optcelectronic components.   Cost (capital investment)
may be traded for labor savings at several possible levels.

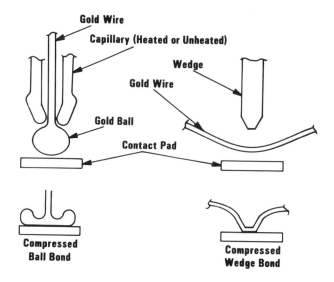

Figure 3.10   Thermocompression ball-and-wedge bonding. Heat and pressure combine to form a metallurgical bond between the contact pad and the gold wire. Ball bonding is used to make the IRED connection, while wedge bonding is used to attach the gold wire to the lead frame post.

made utilizing gold wire and a combination of temperature and pressure to form the electrical contact. Figure 3.10 illustrates this technique.

In both thermocompression ball-and-wedge bonding, the pressure is supplied by either a weighted capillary (dispenser and guide of the gold wire) or a wedge (chisel that deforms the gold wire), and the heat is supplied by heating the substrate and/or the capillary or wedge. The combination of heat and pressure forms a metallurgical bond between the bond wire and the contact pad. Ball bonding is usually the method of contacting the IRED pad, while wedge bonding is usually the method of contacting the header or lead frame post.

In ultrasonic bonding the wire is deformed in a controlled fashion while it is being vibrated at ultrasonic speeds. This vibration creates a metallurgical bond between the wire and substrate. This is similar in appearance to wedge bonding but allows bonding to be performed at lower substrate temperatures.

Thermosonic bonding is a combination of thermocompression and ultrasonic bonding technqiues.

The third operation is the addition of the silicone compound that serves both to increase the amount of photons that escape the

3. IRED Packaging

chip and in plastic packages to cushion the bond area from the shear stresses as the device goes through an increase or decrease in tempperature. Figure 3.11 shows an IRED chip mounted in the well, and with the silicone gel around it.

The fourth operation is the final encapsulation of the device. In metal or hermetic parts this is accomplished by either welding or soldering. The piece parts are shown in Figure 3.12.

In the solder process the lens ring, the solder preform, and the main portion of the package are placed into a graphite boat, which is then heated to a temperature that will allow the solder to flow. The convex lower portion of the lens that will extend into the base of the package allows a self-centering feature. The extra gold bond wire that was wedge bonded to the top of the base secton of the package goes into solution and is absorbed by the solder. The device is now complete and ready for testing.

In the welded enclosure the can containing the glass lens is contained by an electrode that presses against the extrusion at the base of the can. The bottom electrode is beneath the header base. As the electrodes are pressed together the extrusion or lip of the can will scrape the header base creating a low resistance contact area. Current is then passed through the electrodes and the metal package creating heat that merges the metal can to the header base. Examination of the weld area will show smooth metal flow without extrusions. The series resistance between the upper and lower electrodes must be uniform throughout the circumference of the contact to ensure a satisfactory weld. Viewing this area will reveal a smooth and continuous surface of the metal melted from the package and header base. No voids or extruded metal should be visible. The device is now complete and ready for test.

In plastic encapsulated devices the final package is formed by either casting or transfer molding. The cast parts usually utilize molded plastic inserts, which are contained in a machined holder. Mold material is dispensed into each of the 20 cavities and the 20 unit lead frame shown in Figure 3.8 is then inserted. Heat and time cause the cure of this mold compound. The plastic inserts are discarded after several uses. This system leaves a relatively uncontrolled surface at the lead egress point of the package but allows the total lens area and sides of the package to be free of mold marks.

Transfer molding is a similar process in that the lead frame is placed in a holder that has 20 machined cavities. The holder is in two parts, a top section and a bottom section, both of which enclose the unit. The plastic flows under pressure and temperature and subsequently fills the 20 cavities. The holder is removed from the press, the mold halves separated, and the encapsulated devices removed. The separation point of the two sections of the mold is called the *parting line*. Plastic is usually extruded into this interface during the molding process and must subsequently be removed. This excess

The Source (Infrared Emitting Diodes)

Figure 3.11   Detail of the IRED mounted, bonded, and coated with the silicone gel. In addition to the refractive index matching feature of silicone gel, it also serves to "ruggedize" the wire bond against shock, vibration, and thermal cycles.

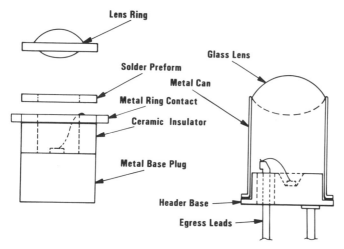

Figure 3.12   Sealing or final encapsulation of hermetic or metal can package. Solder process or welding may be used to create the hermetic seal completing the device. Electrical testing and inspection will take place prior to shipment.

3. IRED Packaging

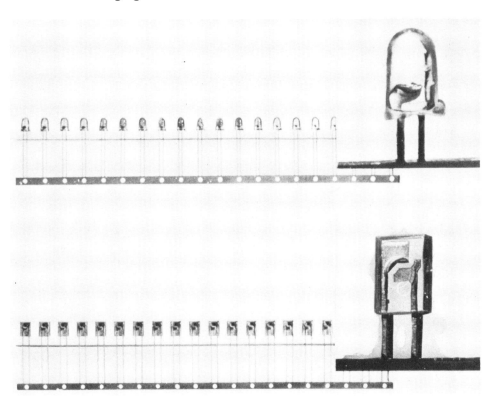

Figure 3.13 Twenty unit lead frames that have been cast and molded. The extra material, or flash, around the molded parts is removed in the next operation, called deflash. The bottom of the cast parts has a meniscus between the egressing lead frame and the edge of the base.

material is called *flash*. The lead frames are then segregated and the devices are ready for testing. Figure 3.13 shows the 20 unit frames after removal from the molding fixture. Note the plastic flashing around the units and lead egress points on the molded parts, and the uncontrolled surface on the bottom of the cast parts.

3.4 CHARACTERIZATION OF THE PACKAGED IRED

This section will cover the characterization of both GaAs and GaAlAs IREDs in various size chips and in different packages. The material will be discussed in both a comparative manner and an explanatory manner. The first portion of the discussion will cover the forward voltage drop of the IRED. When current is passed through the IRED in the forward direction very little current flows until the voltage reaches approximately 0.9 volts. This voltage overcomes the built in potential barrier discussed in Section 1.1 and is known as the diode threshold voltage. The slope of the V_F/I_F curve then changes to a fairly constant logarithmic slope.

Once that voltage drop is exceeded photons begin to be formed. As the current is increased in a well-controlled IRED, the contact resistance on both the N-type and P-type regions will give an increase in voltage drop versus current. Another factor enters into the picture with respect to the N-type regions. Since the N-type contact is smaller than the chip cross section, the resistance from the contact through the N-type region will not linearly increase with increasing current. The total junction must first be turned on and then as the current density increases, it becomes saturated. Figure 3.15 graphically shows this condition. If we assume the chip is 0.016 in. × 0.016 in., then 10 mA of current flow would correspond to 6.0 A/cm^2. This IRED would be reasonably linear in the range of 10 mA to 3 amps. A smaller IRED, such as 0.010 in. × 0.010 in., would be reasonably linear in the range of 4 mA to 1.2 amps. This curve will change in scale with different impurity concentrations. As the impurity concentrations at the IRED are decreased, the curve will shift to the left. Planar diffused IREDs have a current density pattern that shifts this curve to the left due to the lower impurity concentration or doping level.

The forward voltage drop will be reasonably linear in the range of 10 mA to 3 amps on the 0.016 in. × 0.016 in. chip and from 4 mA to 1.2 amps on the 0.010 in. × 0.010 in. chip. At low forward currents (<1 mA) nearly all of the current flow is utilized for surface recombination, tunneling phenomena, space charge recombination, and bulk recombination due to anomalous impurities. As the current increases, these nonideal mechanisms become saturated and become a lower and lower portion of the total current. Conversely as the forward current approaches 2000 A/cm^2, current crowding or saturation tends to reduce the emitting efficiency.

The net effect is to cause the IRED to have a peak operating efficiency at a current that is a function of the geometry of the junction and the size of the electrical contact. This ignores the heating effect caused by the voltage across and the current through the IRED. This factor will be discussed in Section 3.5.

Figure 3.16 shows the forward voltage drop versus forward current for current ranges of 0.1 mA to 100 mA and 0 to 5 amps. These curves are taken from GaAlAs IREDs measuring 0.016 in. × 0.016 in.

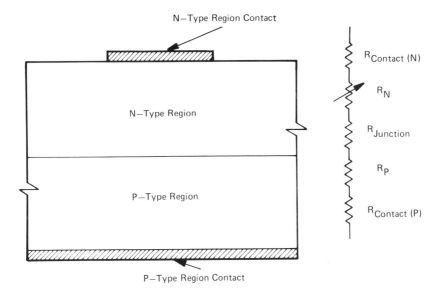

Figure 3.14 IRED cross section with conduction resistances. As conduction occurs in the IRED, the initial voltage drop will equal the voltage required to overcome the diode threshold voltage of approximately 0.9 V. The subsequent slope of the V_F/I_F curve is controlled by the other conduction resistances.

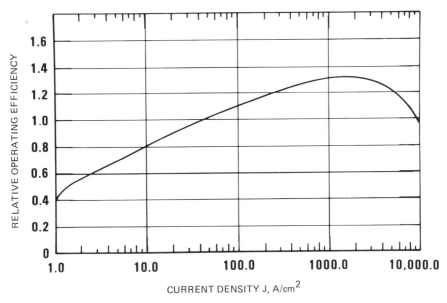

Figure 3.15 Normalized operating efficiency versus current density for an IRED. The curve shown is for solution grown epitaxial IREDs. As the impurity concentration changes, the shape of the curve will remain the same but will shift along the horizontal axis.

These curves will decrease slightly with the change to GaAs and will significantly change with chip size or change in the contact pattern to the N-type side. Note that these curves do not show the effect of heating resulting from the voltage drop across the IRED. This factor is also discussed later in Section 3.5.

The forward voltage drop has a negative temperature coefficient. That is, the voltage drop at a fixed forward current will decrease with increasing temperature. Figure 3.17 graphically shows this change. The units are pulsed so that the device's internal temperature is held close to that of the outside ambient temperature. The plots are for a 0.016 in. × 0.016 in. GaAlAs IRED.

These curves can be used for any IRED by recording the V_F at 25°C ambient temperature at several different current levels, under pulse conditions that minimize junction heating (usually 100 µsec at 0.1% duty cycle) and then sketching a new series of curves "with the same slope" as in Figure 3.17.

The contact to the N-type or top side becomes increasingly critical if the device is to be used under pulse conditions at high currents. A GaAlAs IRED with a 0.016 in. × 0.016 in. cross section is shown in Figure 3.18. The contact area is a compromise that allows V_F to be as low as practical and the output energy to be as high as practical. This is accomplished by making the variable resistance in the N-type region as small as possible (spreading the contact area out in lines) while blocking as little radiating area as possible.

The second portion of this section discusses the emission of the photons from the packaged IRED and presents the advantages and disadvantages of plastic versus metal cans, energy output versus current, and energy output changes brought about by changes in temperature. Table 3.1 shows a comparison of IRED characteristics of plastic and metal packages. Each of the points listed above will be considered in the following discussion.

A cross-sectional drawing of a package with the same dimensional outline in both plastic and metal is shown in Figure 3.19. This package will be referred to throughout this portion of the text to illustrate the points made in Table 3.1.

Package Side Emission

The side of the metal package is a tubular piece of thin nickel or Kovar™ (trade name for a nickel–iron alloy), which is opaque to the infrared energy or photons emitted from the IRED. This material is as smooth as possible and may be plated with a good reflecting material such that energy striking it will be reflected. A large portion of this energy ends up going through the lens and thus becomes useful energy for detection. The sides of the plastic package on the other hand are made of a thermosetting epoxy. If this is not coated with a reflective surface, then any photons that strike this surface at any angle less than the critical angle will penetrate

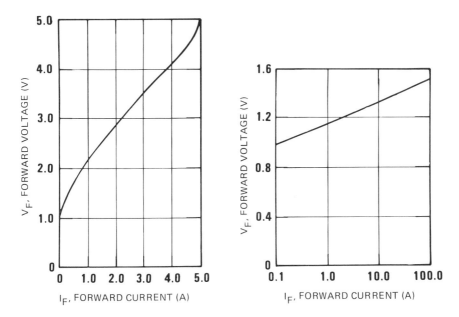

Figure 3.16 Forward voltage versus forward current for 0.016 × 0.016 in. GaAlAs IRED. Changes in material or chip size will alter these characteristics. For example, a change to GaAs would lower the forward voltage. Temperature would also alter these characteristics, as illustrated by Figure 3.17.

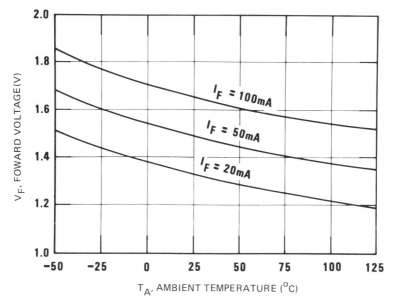

Figure 3.17 Forward voltage versus ambient temperature. Note that diode forward voltage shows a decrease as temperature increases.

The Source (Infrared Emitting Diodes)

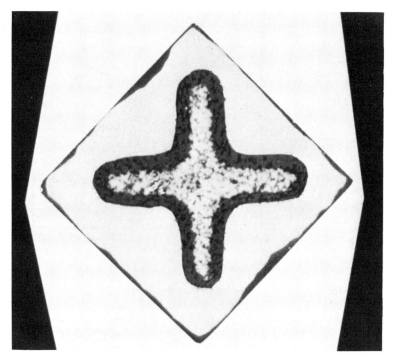

Figure 3.18   N-type region contact on a 0.016 × 0.016 in. GaAlAs IRED chip. Variable resistance caused by the build-up in current density as forward voltage is increased is minimal with this pattern, yet minimum radiating area is blocked.

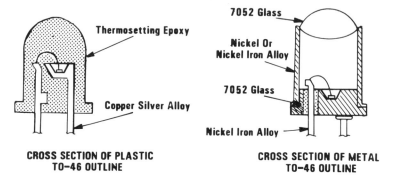

Figure 3.19   Cross-sectional drawing of plastic and metal package. The TO-46 package outline was chosen for the package comparison since it is not only a very popular package type for IREDs but is available in both plastic and metal versions. As a result, the comparison of form, fit, and function can be more easily understood.

3. IRED Packaging

Table 3.1 IRED Package Comparison Between Plastic and Metal Packages

	Plastic	Metal
Side emission	Yes	No
Package height	Taller	Shorter
Lens quality	Usually superior	Usually inferior
Chip centering	Usually superior	Usually inferior
Heat dissipation	Usually superior	Usually inferior
Degradation	Usually superior	Usually inferior
Package cost	Significantly lower	Significantly higher
Storage Temperature	Usually lower	Usually higher
Operating Temperature	Usually equal	Usually equal
Temperature shock	Usually inferior	Usually superior
Solvent resistance	Usually inferior	Usually superior
Hermeticity	Usually inferior	Usually superior
Mechanical shock and vibration	Usually superior	Usually inferior

the package walls and go out the sides of the package. As much as 50% of the radiated energy may be emitted from the sides of the plastic part due to this mechanism. In most cases, this side emission would be useless.

Package Height

The drawings shown for the two packages show identical heights. However, the lens magnification, or focusing ability, is higher on the metal package than on the plastic part. The plastic part requires a flag on the leads (below the IRED) that prevents movement of the lead frame during insertion into the thermosetting epoxy and subsequent curing. As a result the IRED chip must be mounted further away from the base in the plastic package than in its metal counterpart. This simply means that if the exact mechanical outline is required in both packages, then the plastic part will have a wider beam or radiating pattern. If the same optical pattern is required then the plastic package must be taller to accommodate the chip being higher in the plastic package. The plastic package will be taller by this difference in the distance from the chip position to the package bottom. The double-sided lens in the metal package also improves the focusing capability over the single-sided lens in the plastic package.

Lens Quality

The lens on the metal package is usually made from a hard glass (7052) whose thermal coefficient matches that of the nickel-iron alloy (Kovar) package material. In the fabrication of this lens the shape or contour of the upper or lower surface is controlled by a graphite boat holding the molten glass while the other surface contour is controlled by the surface tension of the molten glass. The lens is then flame polished. The optical quality of these lenses is quite good, considering their relatively low cost, but leaves much to be desired when compared to precision ground lenses. Figure 3.20 shows beam pattern plots for both the hermetic and the plastic packages. Note that the included beam angle between the 50% or half power points is about 16° on the metal package and 26° on the plastic package. Since the outline dimensions are designed for an exact mechanical fit, the chip on the plastic package rides higher and as a result gives a wider beam angle.

The radiation pattern measurements for IREDs are performed using a 0.0075 in.-circular aperture in front of a photosensor and moving this aperture at a fixed radius through a 90° arc. Note the higher quality radiating consistency of the plastic package when compared to its metal counterpart. Another difference not apparent in these beam plots is the total spread of the production distribution. Analysis of a large number of both plastic and metal parts shows that a much higher percentage of the distribution will be very similar to those

3. IRED Packaging

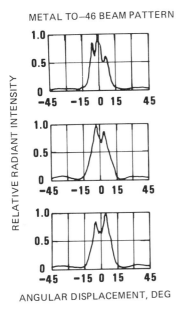
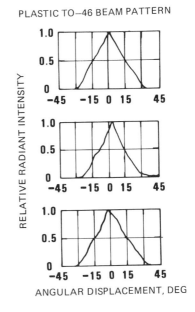

Figure 3.20   Radiation patterns on both the metal and plastic TO-46. Note the wider separation between half power points for the plastic lens. Also note the higher uniformity of intensity for the plastic lens. Beam plots for metal devices do not show as much uniformity from one lot to another as do the plastic devices.

shown for the plastic, while the metal parts will exhibit both more variability in the peak pattern and location.

The beam patterns shown in Figure 3.20 are not typical for all packages. It must be noted that there are a large number of variables that can cause these beam patterns to vary. These include chip placement, chip perpendicularity to mechanical axis, well shape, N_1/N_2 interfaces, lens quality, and distance from the lens tip. Figure 3.21 shows the beam pattern for a narrow radiating pattern in both the metal and plastic TO-46 type package. There are sketched versions rather than a series of actual devices as was shown in Figure 3.20.

The T-1 3/4 is a popular package that first was widely used in visible LEDs and then later used for IREDs. The length/diameter ratio of this package allows a relaively narrow beam angle to be obtained. As the chip location is moved from the top or lens end of the package toward the base, the emitting angle between half power points decreases. Figure 3.22 shows one half of the beam pattern plot as the distance from the lens tip is increased. The beam pattern

Test Conditions: I_F = 100mA
D = 1.429 in. From Flange
Aperture Size 0.0075 in. Dia

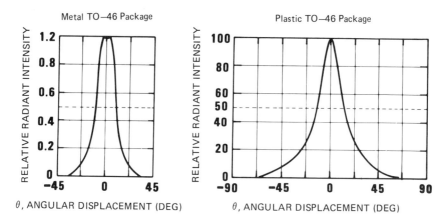

Figure 3.21 Beam pattern for narrow beam plastic and metal TO-46. Note that these are sketched patterns.

Test Conditions: I_F = 100mA DC Aperture Size 0.0075 in. Dia
NARROW ANGLE PLASTIC T-1 ¾

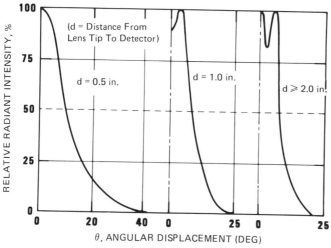

Figure 3.22 Relative radiant intensity versus angular displacement for plastic T-1 3/4 package. Note the changing beam pattern as the distance from the IRED is increased. For this reason, the device cannot be modeled as a point source without error. The beam pattern becomes consistent in shape beyond two inches from the lens tip.

3. IRED Packaging

changes from a single peak to a double peak to a triple peak when this distance is increased from the lens tip to two inches and beyond.

The wide-angle plastic and metal TO-46 are typically substituted for one another based on optical rather than mechanical interchangeability. The plastic package is taller, but the two have similar radiating patterns. Figure 3.23 shows a photograph of each package and the corresponding relative radiant intensity versus angular displacement.

Chip Centering

IRED chip centering is much more a function of manufacturing processing than package variation, because the tapered cup is usually identical in both the plastic and metal package. The lead frame on the plastic part lends itself to automation because the cup in the lead frame for the IRED can mechanically be more easily located. A collet similar in appearance to a drill chuck holds the chip and places it into the cup. The outside of this collet is centered by the tapered sides of the cup giving very precise centering. As a result, the plastic parts, which are usually highly automated, will typically have better centering than their metal counterparts. However, this advantage is partially offset by the difficulty in keeping the lead frame centered within the plastic enclosure while the plastic is curing.

Heat Dissipation

Heat dissipation is usually more efficient for the plastic package if the lead frame is properly designed. The anode lead of the IRED is typically soldered into the land pattern on a double-sided printed circuit board. Since this lead material is usually a copper–silver alloy, the thermal conduction is quite good. The cross section of most anode leads is 0.020 in. × 0.020 in. The metal package on the other hand typically has the lead material made of nickel-iron alloy (Kovar) with a 0.017 in. diameter. This is a much poorer thermal conductor due to the lead material and the smaller cross-sectional area of the leads.

The metal package becomes a better heat dissipator when the side or can portion of the package is mechanically placed in contact with a good heat sink. However, this is rarely done in actual applications due to both added cost and the additional space required. Thermal ratings will be discussed in more detail later in this chapter.

Degradation

The plastic package normally exhibits lower degradation due to the improved thermal resistance discussed in the preceding portion. Figure 3.24 shows a comparison in degradation rate at an I_F (forward

36 The Source (Infrared Emitting Diodes)

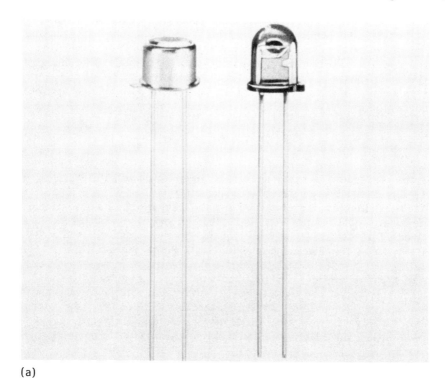

(a)

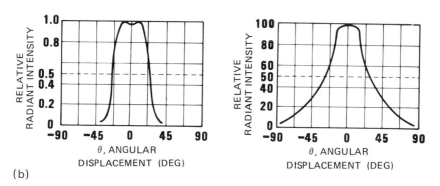

(b)

Figure 3.23 Wide-angle plastic and metal TO-46 photograph and beam angle plots. While mechanical interchange is not possible if fixtures or sockets are utilized, optical characteristics allow easy electrical and optical substitution.

3. IRED Packaging

Test Conditions: $I_{F(DC)} = 100\text{mA}$ At $25°C$
Projected ————————

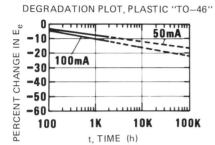
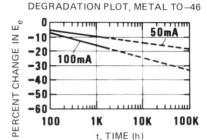

Figure 3.24 Operating life test data on plastic and metal TO-46 packages. The superior output degradation data for the plastic device results from the heat dissipation capability of its silver plated copper leads.

current) of 50 mA and 100 mA for a 0.016 in. × 0.016 in. GaAlAs IRED in both the plastic and metal TO-46 package. This degradation phenomenon is discussed in more detail in Section 1.3.

Package Costs

Package costs were discussed in Section 3.3 on mounting and bonding. A commercial grade GaAs IRED in the T-1 3/4 package in high volume (>500 K/yr) typically costs about one-third of the price of its metal TO-46 equivalent. This cost differential is largely due to the higher degree of automation or added mechanical aids and hence the lower labor costs of the plastic part. At the present time GaAlAs devices are higher in cost due to the increased chip cost and the lower production volumes. As time passes the costs should more nearly equalize.

Storage Temperature Range

The storage temperature range for metal IREDs is usually specified from $-55°$ C to $+150°C$. These limits are somewhat arbitrary but primarily come from limitations specifed for silicon transistors carried over to IREDs. In a silicon phototransistor, h_{FE} (current gain) decreases with decreasing temperature, and I_{CEO} (leakage current) increases with increasing temperature. These two factors lead to the arbitrary limits of $-55°C$ and $150°C$ with h_{FE} controlling the $-55°C$ and leakage current controlling the $150°C$.

The primary stress mechanism in plastic parts is the result of glass transition. This is the temperature at which plastic starts a recure cycle. The stresses that result are caused by thermal expansion mismatches, which can compound or stress the IRED chip or shear the bond contacts or wire. In early plastics utilized in opto components, this glass transition occurred in the 100 to 110°C range. The maximum storage temperature was specified at 85°C range. Improvements in plastic technology have allowed the glass transition rating to be raised to the 125 to 130°C range. As a result, recent device ratings have been raised to 100°C for the package while allowing the chip to be raised to 125°C.

Operating Temperature

The operating temperature usually allows the chip to attain a 125°C maximum temperature. The low thermal conduction of the plastic package keeps the package well below the 125°C danger area. In the future this trend should continue, eventually allowing plastic parts to carry the same storage and operating temperature range as metal parts. At the present time, however, the broader ranges remain with the metal package.

Temperature Shock

Thermal shock follows the same pattern as the storage temperature range. The plastic packages must be kept below the glass transition point or mechanical damage will occur as the plastic goes beyond the glass transition temperature.

Solvent Resistance

Solvent resistance to chemical exposure can be graded into two basic categories. The thermosetting plastics or epoxies that are used in cast parts are less resistant than the temperature/pressure plastics used in transfer molded parts or the metal can parts. The thermosetting materials are not generally harmed by most acids, hydroxides, soaps, and detergents. Exposure to alcohol, gasoline, and most industrial solvents is also nondetrimental. However, acetone and xylene are two common solvents that should be avoided.

For purposes of cleaning or similar short-term exposures, the thermosetting plastic devices can be considered tolerant of any standard chemical that does not show obvious attack on a test sample. For long-term exposures, such as immersed applications, contact the manufacturer for more information.

The temperature/pressure plastics and the metal parts are even more resistant to chemical solvents. In general the two weakest portions of the device are the lead egress points and the marking stamped on the device. If common sense and sample exposure to the particular solvent does not readily supply the correct answer, then contact the manufacturer for more information.

3. IRED Packaging

Hermeticity

Hermeticity is a term that describes the ability of a package to resist the penetration of material from the outside to the inside. The metal packages have an inside cavity and can usually be leak tested by helium or radioactive systems. The plastic packages have no inside cavity and thus must be leak tested in a destructive mode by either a pressure cooker with a steam ambient or by being placed at an elevated temperature in a high humidity environment such as 85/85 (85% relative humidity, 85°C). The seal or leak rate path on the plastic parts is primarily a function of the length of the leak path and the tightness of the bond between the plastic material and the egressing lead. The moisture or other harmful substance must traverse along the lead/plastic interface from the outside of the package to the junction of the IRED. The "cast" plastic parts perform better on hermeticity tests than their molded counterparts. This is due to the reduced internal stress in the cast structure. The small leakage occurring in a nonhermetic IRED is not a big problem, since these devices are operated in the forward mode and increased leakage due to contamination will appear as a very slight reduction in transmitted energy. If the penetrating contaminate is able to attack the IRED chemically, then this argument is not valid. In general, however, the hermeticity advantage rests with the metal package although the plastic package will usually continue to perform adequately.

Mechanical Shock and Vibration

This is usually considered to be a boundary condition for the IRED in a hostile mechanical environment. Impact shock could occur in an application such as a fuse for an artillery shell. The quick acceleration in this application could create severe shock forces. A hostile vibration ambient could occur in an application involving the rotor blades on a helicopter. The plastic package will survive better in these environments because of the containment of the bond wire. Since bond wire movement is *ruggedized*, or constrained, throughout its entire length, the plastic part will have fewer failure modes and thus be more reliable.

Optical Consideration

The beam pattern for a wide beam angle radiating plastic and metal package was shown in Figure 3.23. The corresponding beam patterns for the narrow beam angle radiating plastic and metal package was shown in Figure 3.21. In either the plastic or metal package, if the assumption is made that the chip size and well shape are identical, then the radiated energy should be equal. The wide beam package simply disperses this energy over a wider area. The wide beam angle would be used where the application dictated a wide dispersion of energy. This would occur when the application required an accessory focusing lens such as surveillance (long-range detection) type applications. Figure 3.25 shows the variation in output apertured radiant energy with respect to distance from the lens side of the mounting flange. The aperture used is 0.250 in. diameter and is normalized as shown.

The IRED cannot be treated as a point source of energy when the distance between the IRED package and receiver is short. On the broad beam TO-46 this critical distance is approximately two inches, while on the narrow beam TO-46 the distance becomes approximately six inches. Once these distances are exceeded then the inverse square law of a point source can be used. The inverse square law means that as the distance between the point source and the sensing area increases by a factor of two, then the energy per unit area decreases by a factor of four. This will be discussed in more detail later on in the chapter. Figure 3.26 shows this falloff graphically by illustrating the relative coupling on a production spread of photosensors and IREDs in the TO-18/TO-46 packages. Note the

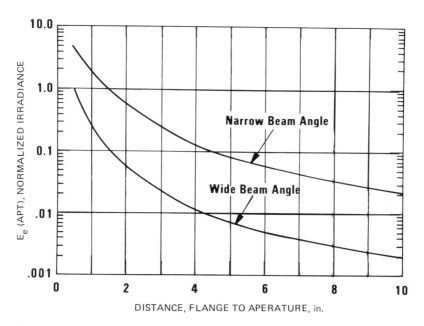

Figure 3.25 Percent change in apertured power output versus distance. Note that the inverse square law does not apply until the sensor is further than 2 in. from the lens on the side beam angle unit and 6 in. on the narrow beam angle unit.

3. IRED Packaging

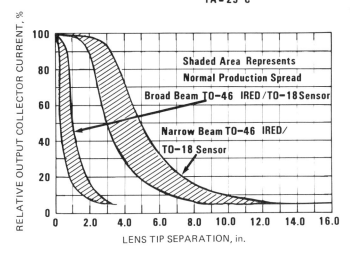

Figure 3.26 Coupling characteristics of broad and narrow beam TO-18 sensors and TO-46 IREDs. The slope of the coupling characteristic curve changes dramatically as the inverse square law takes over.

change in slope on the broad angle parts at approximately two inches and on the narrow angle part at approximately six inches.

The radiant energy of an IRED is not linear versus forward diode current. This was discussed in the first portion of Section 3.4 The output versus current is a function of the metallization geometry of the top or N-type side of the IRED and the cross-sectional area of the chip. At high currents, the shape and location of this N-type metallization are particularly critical. The only major difference between GaAs and GaAlAs is the amount of radiated energy (the GaAlAs has the larger radiated energy). The relative change is the same. Figure 3.27 shows this relative change versus current for a 0.016 in. × 0.016 in. IRED. This change is also independent of the type of lens utilized in the package (the wider the dispersion angle the faster the falloff per unit of distance). Note there is less of a noticeable slope change at two and six inches on the broad and narrow radiating angle parts respectively in Figure 3.26 than in Figure 3.25.

The radiant energy emitted from an IRED has a negative temperature coefficient. The curve shown in Figure 3.28 shows this relative change between −55°C and +150°C. The change is normalized to the reading at 25°C and is the same for GaAs and GaAlAs. Note that forward current is held constant.

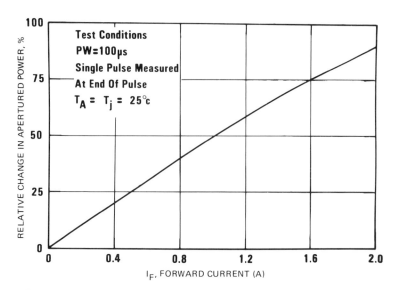

Figure 3.27 Relative change in apertured power versus forward current. Relative changes in emissions as a function of current are independent of package designs but dependent on chip size and metallization geometry.

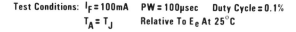
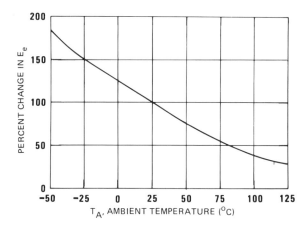

Figure 3.28 Percent change in radiant energy versus ambient temperature. If forward current is held constant, radiant energy will decrease as temperature increases.

3. IRED Packaging

The energy emitted from an IRED is not all of a given wavelength. The IRED emits a band of energy that is centered at a particular wavelength. This center point is near 875 nm for GaAlAs and 930 nm for GaAs at 25°C. Measurement of a number of different production lots shows the center emission at 25°C temperature will vary ±20 nm on the GaAlAs and ±15 nm on the GaAs. This follows logically from the earlier discussion in Chapter 1, which explained the effects of the additional variable of the aluminum in the GaAlAs process. This process requires both more sophisticated equipment and process control. Even with the controls, more variation in the IREDs also occurs. Figure 3.29 shows the spectral emission as a function of wavelength. Note that the GaAlAs emission band extends into the visible red spectrum between 700 and 800 nm.

This spectral emission also shifts with temperature. As the temperature increases, the wavelength of emission gets longer or shifts further into the infrared. Figure 3.30 shows this peak spectral emission shift versus temperature for both GaAlAs and GaAs. Note that the GaAs shifts more than the GaAlAs.

The switching time of the IRED depends on the current density and the material that the IRED is made from. Solution grown epitaxial LEDs switch at slower speeds than the comparable diffused versions of either visible or planar IR types of LEDs. The diffused types have an advantage of faster switching speeds by almost two orders of magnitude, while the solution grown epitaxial devices have a significant advantage in efficiency of photon formation. Figure 3.31 shows the rise and fall times versus forward current for both GaAlAs and

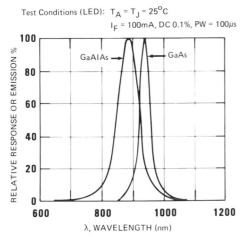

Figure 3.29   Spectral emission versus wavelength for GaAlAs and GaAs IREDs. As discussed in Chapter 1, the GaAlAs epitaxial growth process is more difficult to control, with the concentration of Al being related to the peak wavelength of the finished diode. This explains the wider variance in wavelength for GaAlAs devices.

The Source (Infrared Emitting Diodes)

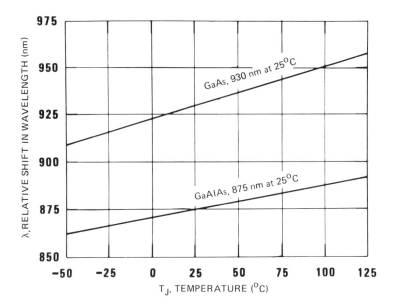

Figure 3.30 Shift in peak wavelength versus junction temperature for GaAlAs and GaAs. The GaAs device peak wavelength will shift more with temperature than the GaAlAs IREDs.

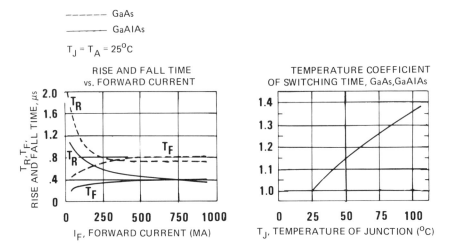

Figure 3.31 Rise and fall time versus forward current and temperature for GaAlAs and GaAs IREDs. Speed is faster for diffused types; however, photon formation is less efficient. Total switching time is faster for GaAlAs than GaAs, decreases with increasing current, and increases with increasing temperature.

3. IRED Packaging

GaAs IREDs. Chip size will have minor effect on these times, causing the speeds to increase slightly with increasing cross-sectional area (increase in capacitance).

3.5 UNDERSTANDING THERMAL IMPEDANCE

The maximum power dissipation rating for a semiconductor device is usually defined as the largest amount of power that can be dissipated by the device without exceeding the safe operating conditions. This quantity of power is a function of:

1. Ambient temperature
2. The maximum junction temperature considered safe for the particular device
3. The increase in junction temperature above ambient temperature per unit of power dissipation for the device package in a given mounting configuration

Item 1 results in lower power dissipation ratings at higher ambient temperatures as described by derating curves, described in the following paragraphs. Item 2 is determined from reliability experiments and is usually considered to be 150°C, although it may be lower due to temperature limits imposed by the package material. Item 3 is called thermal impedance and is determined in the laboratory. The techniques used in this determination are also discussed in the following paragraphs.

Thermal Impedance Calculations

The formula for calculating thermal impedance is

$$R_{THJA} = \frac{T_J - T_A}{P_D}$$

where

R_{THJA} = thermal impedance, junction to ambient (also called θ_{JA}); units are °C/watt

T_J = junction temperature of the device under test

T_A = ambient air temperature

P_D = device power dissipation

R_{THJA} refers to the thermal impedance of a device with no heat sink, suspended in still air on thermally nonconductive leads. This is the worst case (highest value) for thermal impedance.

To calculate the maximum allowable power dissipation ($P_{D(MAX)}$), substitute numbers for R_{THJA} (measured in the lab) and T_J (using the maximum value determined from reliability experiments), then rearrange terms to get

$$P_{D(MAX)} = \frac{T_{J(MAX)} - T_A}{R_{THJA}}$$

This results in a linear power dissipation rating curve which intercepts zero power dissipation at $T_A = T_{J(MAX)}$, and with a slope which is $-1/R_{THJA}$, as shown in Figure 3.32.

The usual (and conservative) method of rating power dissipation is to limit the curve to the safe value for normal room temperature, which is 25°C. The result is a curve shaped like Figure 3.33.

Since there are voltage, current, and ambient temperature limitations not related to chip temperature, the final power dissipation rating curve (often called a derating curve) for a given device will be similar to the curve shown in Figure 3.34.

Since thermal impedance is very nearly constant for different levels of power dissipation, the junction temperature is measured at a known quantity of power dissipation, then substituted into the right side of the formula:

$$R_{THJA} = \frac{T_J - T_A}{P_D}$$

to find the thermal impedance of the device.

It is important to define the ambient conditions, since air movement, lead length, and contact with thermal conductors all affect the measured T_J. The best case (lowest value) of thermal impedance is obtained with an infinite heat sink, i.e., by keeping the entire outside of the device at ambient temperature. Since case temperature equals ambient temperature under these conditions, infinite heat sink thermal impedance is called R_{THJC}, defined as

$$R_{THJC} = \frac{T_J - T_C}{P_D}$$

where T_C = case temperature.

To find R_{THJA} (thermal resistance junction to ambient), R_{THJX} (thermal resistance junction to a finite heat sink), R_{THJC} (thermal resistance junction to an infinite heat sink), the device is placed in the desired mounting configuration and a specific amount of power

3. IRED Packaging   47

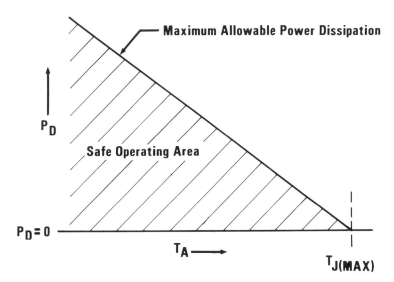

Figure 3.32 Initial thermal derating curve. The curve intercepts zero power dissipation at $T_A = T_{J(MAX)}$ and the slope equals $-1/R_{THJA}$.

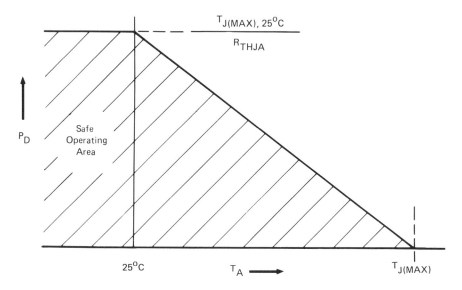

Figure 3.33 Thermal operating curve from 25°C. This curve limits power dissipation to safe values for generation at room temperature ($T_A = 25°C$).

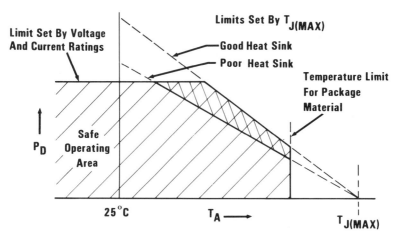

Figure 3.34 Final thermal derating curve. The final thermal derating curve or operating "envelope" is determined by several limits. Voltage and current limitations determine the maximum power dissipation when the ambient temperature is low enough so that no derating is required. The package material is normally limited to operating temperatures lower than the maximum junction temperature.

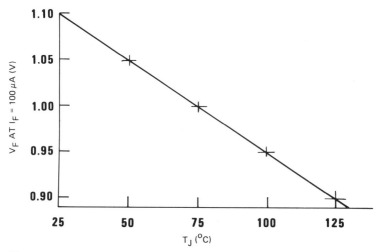

Figure 3.35 Voltage drop versus junction temperature for an IRED. This curve offers the key to accurate measurement of junction temperature at various power levels, necessary for determining proper thermal deratings for IREDs.

3. IRED Packaging

is applied to the device to provide significant chip heating. The junction temperature is monitored by interrupting the power and substituting the low forward bias current (thermometer), 100 µA for the IRED described in Figure 3.35.

The voltage drop must be measured before the junction has time to cool significantly. A 100 µsec interruption is used, which is consistent with the thermal time constant of the devices being measured; a sample and hold circuit maintains the reading so it can be recorded with a voltmeter. The applied waveform for the above IRED would appear as shown in Figure 3.36.

Because of the sample and hold circuit, the voltmeter reading reflects the junction temperature of the chip as shown graphically in Figure 3.35. For a typical plastic package IRED, the junction temperature rises after application of DC power for several minutes, as shown in Figure 3.37.

When the voltmeter reading has stopped changing, (a) substitute the reading back into the graph to get the actual T_J; (b) multiply the large forward current, in this case 100 mA, by the voltage drop on the diode with 100 mA applied to get the power dissipaton; (c) measure the actual T_A; and (d) substitute into the R_{THJA} formula to get a value for thermal impedance.

EXAMPLE: A typical T-1 3/4 packaged 0.016 × 0.016 IRED is as follows.

T_A	V_F (volts, at I_F = 100 µA)
25	1.080
50	1.030
75	0.980
100	0.930

The unit is then connected to a test circuit and immersed in agitated silicone dielectric fluid at a temperature of 25°C. This is a good approximation of an infinite heat sink for a low-power device. An I_F of 100 mA is applied. Every 100 ms the I_F is reduced to 100 µA for a period of 100 µsec, after which the I_F returns to 100 mA (see Figure 3.36).

Using a sample and hold circuit, it is observed that the V_F of the device during the low current intervals starts out at 1.080 V but rapidly decreases, eventually stabilizing at 1.050 V. Interpolating

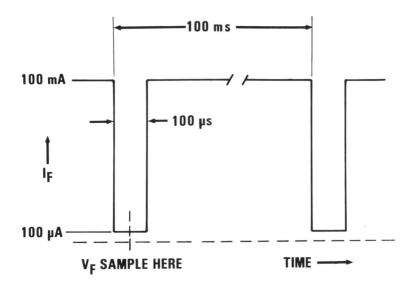

Figure 3.36 Timing cycle for device heating and monitoring of junction temperature. The objective is to measure junction temperature quickly by taking a voltage drop reading, while having minimal interruption of the power level being tested.

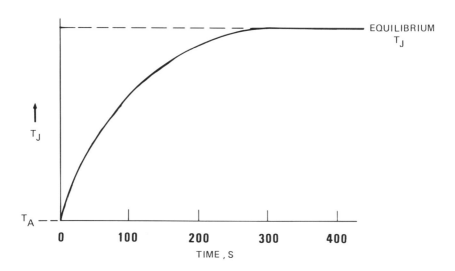

Figure 3.37 Equilibrium of junction temperature. The application of DC power causes a rise in junction temperature, which tends towards a new equilibrium, or steady state, temperature.

3. IRED Packaging

between 1.080 V (25°C) and 1.030 V (50°C), it is found that the junction temperature is now 40°C.

The V_F is measured during the 100 mA I_F period and found to be 1.50 V. Thus, the power dissipation is 150 mW (99.9% of the time). Substituting into the formula,

$$R_{THJA} \text{ (infinite heat sink)} = R_{THJC} = \frac{40 - 25}{0.150} = 100°C/W$$

When the same test is conducted with the device in still air, mounted in a PC board socket, the final values of V_F are 1.024 at 100 µA and 1.40 at 100 mA. Thus $T_J = 53°C$ and

$$R_{THJX} = \frac{53 - 25}{0.140} = 200°C/W$$

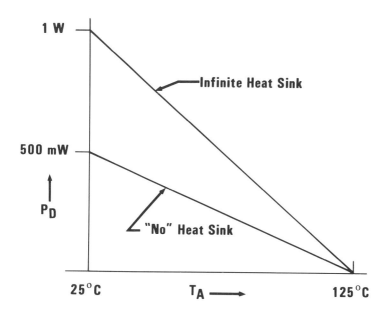

Figure 3.38 Thermal derating for "infinite" and "no" heat sink. With no heat sink, logic predicts that the P_O rating will be lower; hence, the bottom line represents operation with no heat sink. In practice, the optimal derating would lie somewhere between the two curves.

The power derating curves are:

$$P_D = \frac{T_{J(MAX)} - T_A}{R_{THJA}} = \frac{125 - T_A}{100}$$

with infinite heat sink, and

$$P_D = \frac{125 - T_A}{200}$$

with no heat sink.

Graphing the derating curve gives two lines, as shown in Figure 3.38.

Due to the fact that the plastic package can withstand only 100°C (because of the glass transition temperature), the device is limited to 250 mW. Thus, the final power derating curve is shown in Figure 3.39.

The entire shaded area can be used with an infinite heat sink; the cross-hatched area is forbidden for a device with no heat sink.

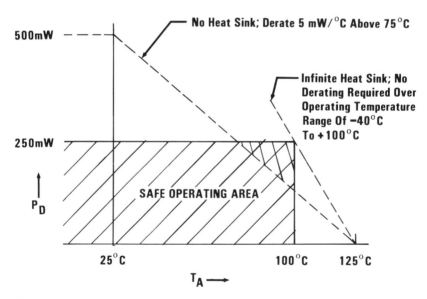

Figure 3.39 Final thermal derating. Only a small portion of safe operating area extends into the region between zero and infinite heat sinking. The vast majority of applications will facilitate operation in the portion below the line representing no heat sink.

3. IRED Packaging

3.6 UNDERSTANDING THE MEASUREMENT OF RADIANT ENERGY

Infrared emitting diode power measurement depends on a number of variables that must be precisely defined in order for design engineers to utilize manufacturers' data sheet information. Manufacturers differ in the techniques used in measuring power and also in their interpretations of the definitions of measured parameters. This section should help clarify the differences, especially those related to GaAs and GaAlAs solution grown epitaxial devices.

General Discussion

Power is measured in units of energy per unit of time, and the conventional MKS unit is the watt. Some factors that must be controlled to make accurate power measurements are discussed below.

The energy an IRED emits is in the form of photons, and the photon's energy is inversely proportional to its wavelength. To measure the power emitted, the technique must take into account both the rate of photon emission and the average wavelength of the emitted photons. Both of these vary as functions of chip temperature. See Figures 3.40 and 3.41 for examples of these changes.

Stress on the chip will cause any defects in the chip to expand along the planes of the crystalline structure in a process called dark line defect formation. This degrades the chip, causing the power output to decrease. These dark line defects effectively destroy the crystalline formation and thus reduce the ability to create photons. Throughout the chip's operating life, these dark line defects continue to propagate, resulting in time-dependent output degradation. Measurements made after the chip has been stressed mechanically, thermally, or electrically will be lower than initial readings. If the operating and environmental elements are accounted for, then the IRED degradation becomes a predictable function of time. Many manufacrurers provide information on output reduction versus time for fixed current, temperature, and heat sink conditions. Figures 3.42, 3.43, and 3.44 illustrate the magnitude of these output changes due to applied DC current for variations of ambient temperature, current level, and different materials used as emitters.

The response of most detectors is also wavelength and temperature dependent. One reason for these variations is that the surface of the detector can reflect photons depending on the wavelength, the angle of incidence, and the type of protective coating on the surface. The range of linearity in power detection can also be exceeded by some emitting devices. Finally, there are other minor characteristics of detectors to be considered. Obviously, the accuracy of the detection system is critically important to measuring the output of an IRED accurately.

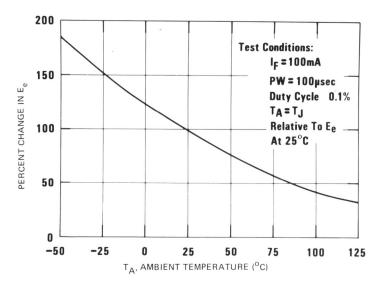

Figure 3.40 Output power versus ambient temperature for both GaAs and GaAlAs IREDs. Output decreases as temperature increases for these devices.

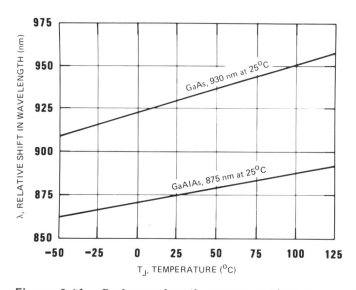

Figure 3.41 Peak wavelength versus ambient temperature for both GaAs and GaAlAs IREDs. Peak wavelength increases as temperature increases for these devices.

3. IRED Packaging

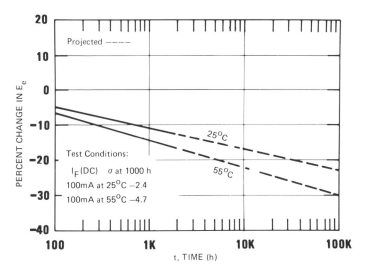

Figure 3.42 Percent change in GaAs IRED mounted in metal TO-46 package versus time at 25°C and 55°C. Note the increase in degradation caused by the higher ambient temperature.

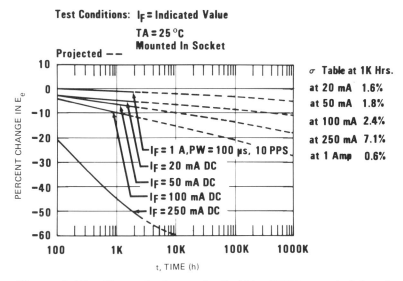

Figure 3.43 Percent change in GaAlAs IRED mounted in plastic TO-46 package versus time at various current levels. Note the increased rate of degradation caused by higher current.

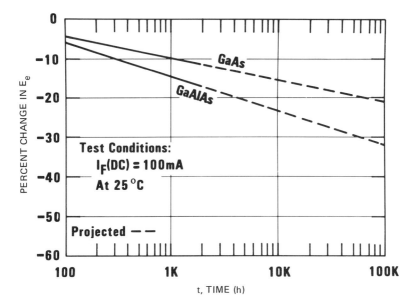

Figure 3.44 Percent change in GaAs and GaAlAs IRED mounted in metal TO-46 package versus time under same conditions. Note the increased susceptibility to degradation effects of GaAlAs under the same forward current and temperature conditions. Remember that the GaAlAs is emitting more usable energy than the GaAs for the same current. The peak wavelength of the GaAlAs IRED is also more closely matched to photosensor peak spectral response.

Any measurement of directed output is dependent on complex optics, which include chip centering in the reflective cup, reflector design, chip-to-lens centering, bubbles or energy blocking contaminants in the packaging, and the fact that less than half (40%) of the emitted photons exit the chip from the side walls rather than the top surface.

Many devices have radiation patterns that change as the distance from the device to the detector is varied, so this distance can be important in directed output measurement. Logically, this distance also becomes critical in slotted switch design, as discussed in Chapter 8 (see Figure 3.45).

3. IRED Packaging

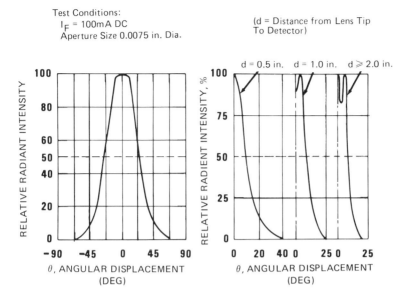

Figure 3.45 Relative radiant intensity versus angular displacement for T-1 3/4 package. The *beam angle* is generally measured between the half power points and may be expressed as the included angle (from one half power point to the other) or the half angle (from one half power point to the optical axis or zero angular displacement line), depending on the manufacturer's specification method.

Parameter Definitions and Measurement Techniques

There have traditionally been two methods of defining power measurement, but there have been different interpretations for each.

The first method is radiant power output (P_O or E_e), sometimes called total power. A strict interpretation of P_O is that the total amount of radiation exiting the package, regardless of direction, should be measured. Some manufacturers have interpreted radiant power output to be only that radiation that exits the package in a direction useful to most designers.

The measurement may include only that radiation collected by a flat surface detector near the lens tip and orthogonal to the lens axis. Radiation emitted from the sides or back of the package and surface reflections from the detector are not collected. Therefore, the devices from these manufacturers are conservatively rated (sometimes by as much as a factor of two, depending on the device type). Their output may actually be quite high when compared to devices measured differently by other manufacturers. For instance, P_O readings for the narrow (15° between half power points) radiation pattern plastic packages utilizing the 0.016 in. × 0.016 in. GaAlAs IRED are typically 60% higher when using a parabolic reflector than when using a flat P_O test fixture. This is due primarily to the collection of radiation from the sides of the package. P_O measurements are normally useful only for devices with wide radiation patterns, because the primary application is in providing a relatively even intensity over a large area. Radiation that exits the side or back of the package is not useful without external reflectors; and if external reflectors are added, there are intensity peaks in the radiation pattern that are detrimental in most applications.

The second way to measure power is on-axis intensity. This is done by measuring the power incident on a specified area. The most common method is to utilize a fixture that controls the distance from the device to a measured aperture on the detector. This measured power can then be specified as average power per unit area (both $E_{e(APT)}$ and P_A are equivalent and the measurement is usually expressed as mW/cm^2) or as I_e average power per unit of solid angle (i.e., where the measure is expressed as mW/sr or milliwatts per steradian). Note that measurements expressed in mW/cm^2 should also include information on the distance from the device to the aperture and the size of the aperture, in order to be useful to design engineers.

The calculated value of I_e is also dependent on distance for most applications, and a design engineer can be misled by the mathematical model into assuming that I_e is a constant regardless of distance. Most IREDs cannot be modeled as a point or discrete source except at distances that are very large compared to the package dimensions and/or optical dimensions. Thus, the foundation assumption in spherical calculations (using mW/sr) is invalid and attempts to use this model may lead to errors. Note in Figure 3.46 how the mW/sr becomes consistent after approximately six inches separation.

Some manufacturers have chosen to use $E_{e(APT)}$ or P_A rather than I_e for devices that do not have a virtual source that is distance independent. This is the preferred parameter because a simple performance graph can then show how $E_{e(APT)}$ varies with distance, as shown in Figure 3.47.

$E_{e(APT)}$ measurements have historically been made only for narrow radiation pattern devices because their major application is to have a high on-axis intensity for good coupling efficiency with a small sensing area photodetector (see Figure 3.48).

However, some manufacturers are beginning to use the measurement parameter with wide radiation pattern devices also. $E_{e(APT)}$

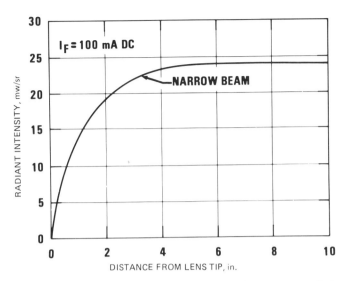

Figure 3.46 Output intensity in mW/sr versus distance from lens tip on T-1 3/4 package. A consistent function of separation distance only occurs after approximately six inches; therefore, use of geometrical models to predict irradiance can lead to design errors at close proximity.

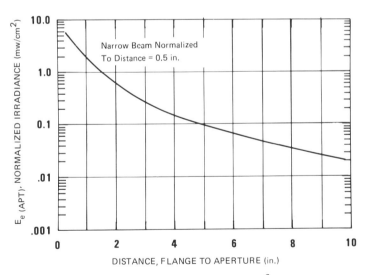

Figure 3.47 Output intensity in mW/cm^2 versus distance from lens side of mount area. This method of specifying device output is convenient for design engineers, because most photosensors are tested and specified in terms of radiant intensity expressed in mW/cm^2, making sensor output current prediction an easy calculation.

The Source (Infrared Emitting Diodes)

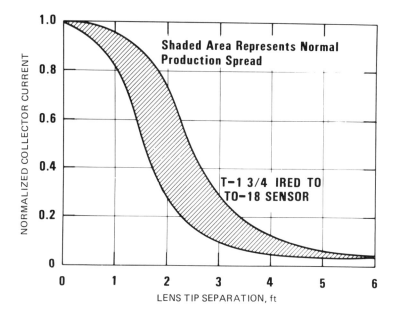

Figure 3.48 Coupling characteristics of plastic TO-18 phototransistor and GaAlAs IRED versus separation between lens tips. Virtually all of the complex optical calculations can be avoided by the use of simple coupling characteristic curves provided by many manufacturers.

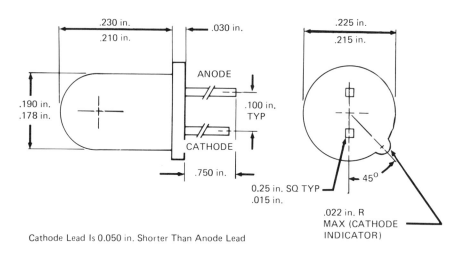

Figure 3.49 Outline drawing from a plastic T-1 3/4 GaAlAs data sheet.

3. IRED Packaging

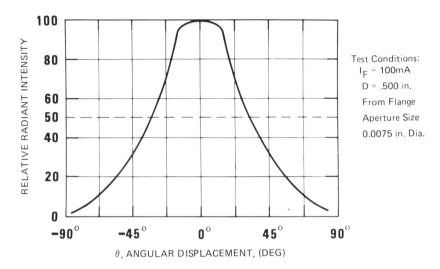

Figure 3.50   Wide-angle beam pattern from a plastic T-1 3/4 GaAlAs data sheet.

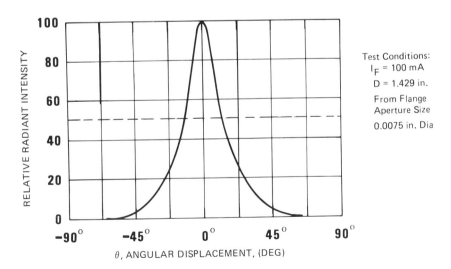

Figure 3.51   Beam pattern of narrow-angle beam pattern from a plastic T-1 3/4 GaAlAs data sheet.

is a key design parameter when the distance and aperture are chosen to give maximum useful information. The distance is normally chosen so two criteria are met: (a) all intensity peaks should fall within the aperture opening for devices with normal optics and (b) the distance should be at a maximum with the constraint that the intensity does not vary more than 10% from point to point within the aperture opening for normal devices. This gives the maximum useful information to the systems designer. Aperture size is typically chosen so that it is slightly larger than the lens diameter of a detector, which is mechanically matched to the dimensions of the IRED. This provides the user with a mechanical alignment tolerance as well as the average power intensity within the aperture.

Conclusion

Power measurement of IREDs varies more than any other parameter among different manufacturers. Part of the difference is in interpretation of the definitions of the parameters measured, and part is in the technique used.

3.7 RELIABILITY

In optoelectronic technology, the two main reliability considerations are long-term IRED degradation and catastrophic failure due to thermal and/or mechanical stress. Stress on the chip will cause any crystalline defects in the chip to expand along the planes of the crystalline structure in a process called *dark line defect formation*. This causes a degradation of power output as these crystalline defects result in a reduction in photon formation. As these defects expand the total output for the chip is measurably decreased. Measurements made after the chip has been stressed will be lower. Once the chip has been mounted and encased by the manufacturer, the primary degradation is controlled by those stresses created by current density, temperature, and time. Most manufacturers will accumulate data on different package styles and different chips under varying stress levels. This data is then plotted, showing a linear decrease in relative output when time is plotted logarithmically. Some manufacturers feel that there is an annealing effect on the defect formation and that this linear line will become asymptotic to a fixed degradation percentage at some time. Since this time period is too long to be tested empirically, the information is presented as a linear decrease versus logarithmic time in order to offer a conservative approach for the systems designer.

Initially (the first 120 hrs), degradation data is very sporadic. Some units improve, some units remain flat, but most units will decrease at varying rates. Once these initial variations are complete,

3. IRED Packaging

the devices will assume a linear decrease. The observed data is usually plotted as an average of the distribution. Sigma, or the standard deviation, is then calculated for each measurement point. The one sigma, two sigma, or three sigma plots can be added, and the system designer can then calculate the degradation rate for the system stress conditions.

One of the methods of minimizing this early variation is to pre-age or burn-in the IREDs. There are two important points to be considered. If the system design is so critical that a pre-age or burn-in is required, then the length of time needs to be established. Twenty-four hours will remove most of the variation; 96 hours will essentially remove all of it. If the system is operated for testing or stabilization then the burn-in may be accomplished during this time. It is almost always bad to pre-stress, pre-age, or burn-in the device under accelerated conditions. Due to the normal way the current distribution in the IRED behaves, an over stress at system operating levels occurs when accelerated aging is employed. This simply means that if the device is to be operated at some lower forward current than the level used for burn-in, then the emitted energy will be abnormally degraded at the lower forward current. For example, if the unit is to be operated at 10 mA and aged at 50 mA, the percentage drop in energy output measured at 50 mA test condition will be lower than that measured using a 10 mA test condition. The accelerated burn-in destroys a useful portion of the 10 mA performance of the device.

Use of thermal derating curves requires an understanding of thermal properties as well as a starting point. The thermal derating implies that the stress level is a linear derating function from 25°C (as temperature is increased) to the temperature at which the stress level is zero. For ease of calculation, we will assume that V_F is constant. If a device is rated for 150 mW at 25°C and V_F at 100 mA is 1.5 V, then I_F maximum is 100 mA. If the device derates to zero at 125°C then the device will be rated for 100 mA at 25°C, 75 mA at 55°C, 50 mA at 75°C, etc. The 100 mA degradation curve at 25°C would be identical to the 75 mA degradation curve at 50°C, which would be identical to the 50 mA at 75°C, etc. This also is a conservative approach since V_F is decreasing with both decreasing current and increasing temperature. In fact, a unit operated in the above example will exhibit improved degradation rates as the temperature is increased.

The net result is that a family of curves for various forward currents at a 25°C ambient with associated sigma is all that is necessary for the designer to be able to calculate the system degradtion rate. Such a family of curves is shown for a 0.016 in. × 0.016 in. GaAlAs IRED in a plastic T-1 3/4 package in Figure 3.52.

The curves shown in Figures 3.52, 3.53, and 3.54 show that in-

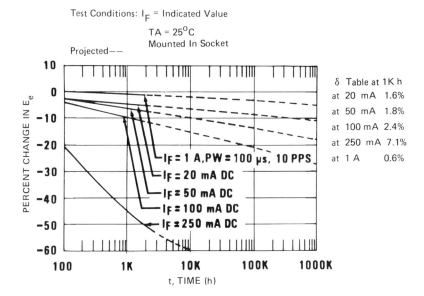

Figure 3.52 Percent change in power output versus time for 0.016 in. × 0.016 in. GaAlAs IRED in plastic T-1 3/4 package. Note that operation at higher forward current will increase the rate of degradation.

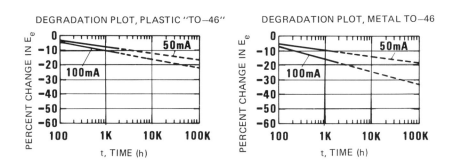

Figure 3.53 Percent change in GaAs IRED mounted in metal TO-46 package and plastic TO-46 package versus time.

3. IRED Packaging

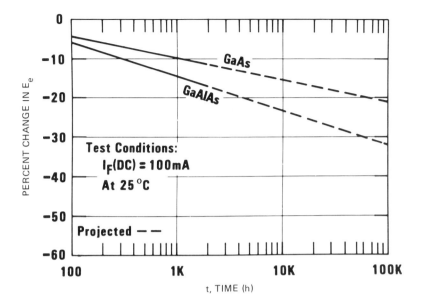

Figure 3.54 Percent change in GaAs and GaAlAs IRED mounted in metal TO-46 package versus time under same stress conditions. Note that GaAs shows less degradation than GaAlAs at identical forward currents.

creasing current causes a faster degradation rate, increasing temperature causes a faster degradation rate, and GaAs degrades more slowly than GaAlAs.

Catastrophic failure due to thermal or mechanical stress results from forces on the IRED chip or wire bond that in turn shear or break the bond wire. If such failures occur, they usually occur early in the operating life of a device. Since the machine fabrication of the plastic part is very repeatable and mechanically accurate, there is a lower probability of obtaining a weak bond. Machine fabrication of plastic devices tends to offset the low shear stress environment of the air chamber inside a metal package. The net effect is equal reliability. Catastrophic failures are very infrequent in plastic and metal package styles as long as production is properly controlled by the manufacturer. As a result, neither plastic nor metal can be said to be less reliable than the other. The greater variation is found among the different manufacturers.

II
THE RECEIVER (SILICON PHOTOSENSOR)

4
The Photodiode

4.1 BASIC THEORY

The formation of a P-N junction in silicon and the corresponding built-in potential barrier is identical to the discussion in Section 1.2 for IREDs. However, in the receiver or silicon P-N photodiode, the main objective is the generation of photocurrent.

Semiconductor material will generate a photocurrent on absorption of incident radiation. The photoelectric effect takes place within the semiconductor as photons excite electrons and "kick" them from the valence band into the conduction band. A hole in the valence band is left for each addition to the conduction band of "free" electrons. Photo excitation in P-type material results in the creation of additional minority carriers (electrons), which generate a photocurrent on crossing the P-N junction.

If the semiconductor does not have an externally applied voltage, these minority carriers must reach the P-N junction by diffusion. If the electron hole pair is formed more than a diffusion length from the junction, simple recombination is likely to occur with no contribution to photocurrent.

Response time for a device is relatively slow when there is no electrical field present to accelerate newly formed minority carriers toward the P-N junction. This mode of operation is known as *photovoltaic operation*. When an external bias is applied to the semiconductor, the response time speeds up dramatically with the addition of a depletion region near the junction. This region has a field gradient sufficient to accelerate the newly formed electrons, across the junction. This type of operation is referred to as the photoconductive mode. Figure 4.1 shows the electric field with the recombination of electron-hole pairs and the separation leading to photocurrent.

Analysis of Figure 4.1 shows that the electric field is not uniform. The field is much stronger in the depletion region than it is in the adjacent N-type or P-type regions. The field shown applies to the diffused junction shown in Figure 4.2.

Figure 4.2 shows the depletion region in a diffused structure with respect to penetration into the N-type starting material and the P-type diffused region. It should be remembered that in the P-type

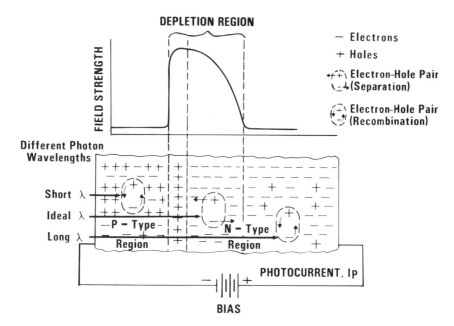

Figure 4.1 P-N photodiode junction—photocurrent created within internal field. Photons traverse the crystal a distance that is dependent on their wavelength. Electron-hole pairs that are far away from the junction or not generated in the presense of a strong field usually recombine before a photocurrent can be generated.

region the majority carriers are holes and the minority carriers are electrons. The N-type region is the inverse where the majority carriers are electrons and the minority carriers are holes.

As the reverse voltage across the P-N junction is increased, the depletion region will expand. For a given voltage, the expansion into the P- or N-type regions is governed by the minority carriers within that region. The depletion region has to deplete the same number of minority carriers from the N-type region as are depleted from the P-type region for a given voltage. The number of minority carriers within the P-type region is significantly higher than the number within the N-type region. Therefore, the depletion region is wider in the N-type region than in the P-type region.

The depletion region exists even with zero applied bias. The "built-in" field discussed in Section 1.2 is the same for the P-N junction of a silicon photodiode. Reverse bias aids this built-in field and thus expands the depletion region.

A photodiode has the best performance when the largest number of photons are absorbed in the depletion region. An ideal photodiode

4. The Photodiode

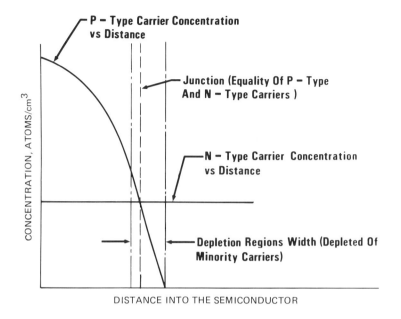

Figure 4.2 Depletion region in a diffused P-N junction. The wider depletion region in the N-type material results from the fact that N-type material has a lower concentration of minority carriers than P-type material. Equal numbers of minority carriers must be depleted on either side of the junction in order to build an electrical potential equal to the reverse bias voltage.

would have a depletion region wide enough to allow absorption of the desired photon wavelengths within this depletion region. The depth which a photon travels before it is absorbed is a function of its wavelength. Short photon wavelengths (such as ultraviolet or visible energy) are absorbed near the surface, while those having longer wavelength (such as infrared) may penetrate the entire thickness of the semiconductor crystal.

Optimization of "capacitance for speed" and "depletion region width for conversion efficiency" would yield a PIN diode structure. This provides the most conversion efficiency possible for a commercial device. The starting material is as close to intrinsic silicon (having a minimum amount of majority carriers) with very little compensation (having a minimum amount of minority carriers) as possible and a very shallow N-type diffusion on one side with a corresponding shallow P-type diffusion on the other side.

This approach yields a device with low contact resistance, high conversion efficiency (photons creating hole-electron pairs within the depletion region), and correspondng low capacitance per unit area. Most manufacturers purchase a silicon slice of near intrinsic resistivity (either P-type or N-type) and make shallow N-type and P-type diffusions into either side prior to metallization for contacts.

4.2 CHARACTERIZATION

Operation of the PN or the PIN photodiode may either be in the forward or the reverse mode. Operation in the forward mode with zero bias is called the photovoltaic mode. The photons impinging on the photodiode generate a voltage potential. Common applications using this mode employ a PN photodiode operated as a voltage source, also commonly called a solar cell. Operation with reverse bias or in the reverse mode is called the photoconductive or photocurrent mode.

When comparing the photovoltaic mode to the photoconductive mode the photoconductive mode offers the following advantages:

1. Higher speed.
2. Improved stability.
3. Larger dynamic range of operation.
4. Lower temperature coefficient.
5. Improved long wavelength response within the planar diffused region.
6. Response to the short wavelengths in the edge region adjacent to the planar diffusion. (The formation of the depletion region at the surface allows hole-electron pairs to be created from the short wavelengths in the depletion region rather than in the N-type or P-type regions.)

Figure 4.3 shows the relative response versus wavelength for a PIN photodiode made with starting material of 1000 Ω per centimeter N-type material or a PN photodiode made with 10 Ω per centimeter N-type material. The geometry of the devices are identical.

Note that the PN photodiode peak relative response is at a shorter wavelength. This is due to the narrower depletion region located closer to the surface and hence the loss of useful hole electron pairs created by the longer wavelength photons. Overlaying this with the GaAs and GaAlAs spectral emission versus wavelength plots in Figure 3.29 clearly shows the better compatibility of the PIN photodiode to GaAs and GaAlAs IRED emission.

The total capacitance versus reverse bias voltages on the same two photodiodes is shown in Figure 4.4. Note the improvement in capacitance inherent in the PIN photodiode. This sharply illustrates the wider depletion region for a given voltage across the PIN structure and hence the lower capacitance. The plates of the capacitors are simply the cross-sectional area of the P-type and N-type regions adjacent to the depletion region with the width of the depletion region being the separation or distance between the plates.

4. The Photodiode

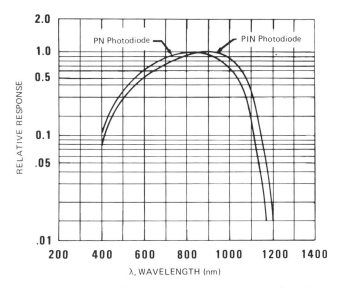

Figure 4.3 Relative response versus wavelength. The wider depletion region of the PIN photodiode results in longer wavelengths being absorbed within the region and forming useful electron-hole pairs. The net result is a relative response peak at a longer wavelength than that of a typical PN photodiode.

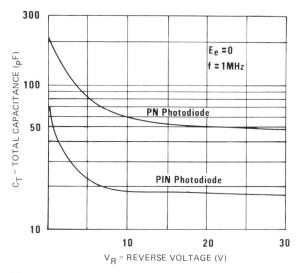

Figure 4.4 Total capacitance versus reverse bias voltage. The physical separation of charge layers at the edges of the depletion region is significantly greater for the PN photodiode. This results in a lower capacitance, making a higher operating speed possible.

The low capacitance of the photodiode when operated in the depleted mode allows the unit to operate at extremely high speeds. The speed of response for a given PIN photodiode depends on the areas of the photodiode that are irradiated, the amount of reverse bias applied, the capacitance, and the value of the effective load resistance.

Photocurrent begins to flow within a few picoseconds after the photons impinge on the photodiode, but there is junction, package, and stray wiring capacitance to be charged. This then makes the rise or fall time largely dependent on the load resistance unless the load resistance is so small that the internal resistance of the photodiode limits the speed. When effective load resistance approaches 100 ohms then this effect becomes pronounced. A PIN photodiode with effective load resistance in the 100 to 500 Ω range should switch in the 0.5 ns to 5 ns range for both rise and fall time. Since the normally available IREDs discussed in Part I will operate only up to a 1 MHz rate, it is readily seen that almost all PN and PIN photodiodes are more than adequate to use as receivers. Figure 4.5 shows a recommended amplifier circuit for a photodiode operated in a linear mode.

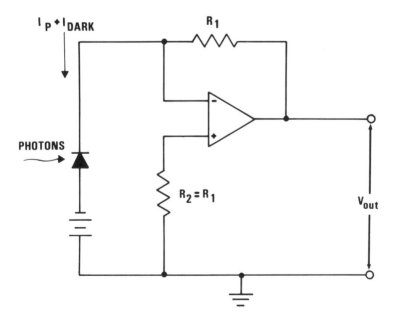

Figure 4.5 Linear mode circuit for photodiode and amplifier. While offering the advantage of fast rise and fall times, photodiodes almost always require the addition of an amplifer to be useful as infrared detectors.

4. The Photodiode

The negative-going input is very close to ground potential. The dynamic resistance seen at this negative-going input by the photodiode is R_1 divided by the loop gain. If the operational amplifier has extremely high input resistance, then the loop gain closely approximates the forward gain of the operational amplifier. This type of application under open air communications will be discussed in more detail in Part V.

Another major advantage of the PN or PIN photodiode is the linearity of response with respect to the quantity of photons impinging on the surface. These photodiodes can effectively be used from photon levels that will produce photocurrent slightly above the dark current level (50 to 100 picoamps or 50×10^{-12} amps) to direct sunlight in the 80 mW/cm² level. This corresponds to approximately nine order of magnitude. Figure 4.6 shows this linearity graphically for both a PN or PIN photodiode.

The photocurrent of a PN or PIN photodiode depends on a number of variables. Ideally each photon (quantum of energy) should cause one electron to be added to the stream of photocurrent. "Quantum efficiency" is therefore dimensioned as "electrons per photon." Most manufacturers express this in terms of the *flux responsivity* (R_ϕ). This takes into account the photon energy, and represents the ratio of photocurrent to the amount of spot flux:

$$R_\phi = n_q \left(\frac{\lambda}{1240} \right) = \frac{I_p}{\phi_e}$$

where

R_ϕ = flux responsivity in amps per watt

I_p = photocurrent in amperes

ϕ_e = radiant flux in watts

n_q = quantum efficiency in electrons per photon

λ = photon wavelength in nanometers

R_ϕ will approximately follow the curve shown in Figure 4.3, where the relative response that is normalized to 1.0 is changed to flux responsivity with a maximum of 0.60 amps/W. This is shown in Figure 4.7.

The quantum efficiency on the PIN photodiode at 900 nm with R_ϕ equal to 0.6 A/W is:

$$0.6 = n_q \left(\frac{900}{1240} \right)$$

$$n_q = 0.83$$

$$n_q = 83\%$$

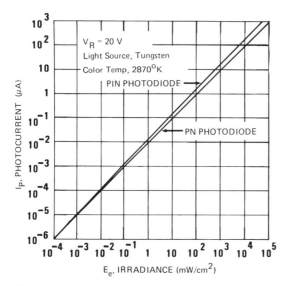

Figure 4.6 Photo current versus irradiance for a PN and a PIN photodiode. Linearity over nine decades is achievable for photodiodes. Irradiance levels range from virtually dark to sunlight (10^2) and higher.

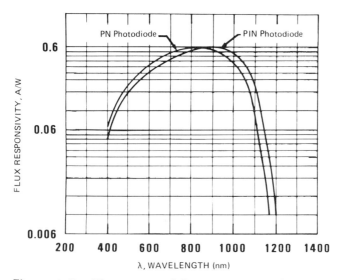

Figure 4.7 Flux responsivity versus wavelength. Flux responsivity is analogous to the relative spectral response. Controlling variables include the width of the depletion layer itself and the distance from the surface of the crystal to the depletion layer.

4. The Photodiode

Another measure of performance is the incidence response, R_E, which adds the variable of the photosensitive area (or the magnified area if a lens is used). This is equal to the ratio of photocurrent to incidence:

$$R_E = \frac{I_p}{E_e} = \int [R_\phi(A_D)] \, dA_D \sim R_\phi A_D$$

R_E = incidence response in amps/watts/square centimeters

E_e = radiant incidence in watts/square centimeters

A_D = effective photosensitive area in square centimeters

I_p, R_ϕ (from previous equation)

Assume a radiant incidence of 2 mW/cm^2 at 900 nm (peak response) and a diode effective area of 0.065 cm^2 ($\sim$0.100 in. × 0.100 in.)

$$I_p \sim E_e R_\phi A_D$$
$$\sim 2 \text{ mW/cm}^2 \, (0.6 \text{ A/W}) \, 0.065 \text{ cm}^2$$
$$\sim 0.13 \times 10^{-3} \, (0.6 \text{ A/W})$$
$$\sim 78 \mu A$$

(assumes radiant incidence at 900 nm or peak response)

There is noise associated with operation of the photodiode. The two types of noise to be considered are *flicker* noise and *shot* noise. The two components of current that contribute to this noise are junction current and leakage current.

The junction current causes *shot* noise while the leakage or dark current causes thermal noise from the leakage resistance and *flicker* noise. A worst case value of the leakage current noise is obtained by applying the *shot* noise formula to the entire dark current. This thermal noise is usually dominant below 20 Hz.

When the frequency is increased to greater than 20 Hz the *shot* noise becomes dominant. The signal-to-noise ratio is defined as the ratio of the photocurrent, when signal is applied, to the noise current, when there is no signal. This signal-to-noise ratio increases as the square root of the photodiode area, because the signal rises linearly with the area while noise current varies as the square root of the area. The noise equivalent power, or NEP, is defined as the signal flux level when the signal-to-noise ratio is one and the band width is narrow (1 to 10 Hz). The NEP varies inversely as the responsivity (see Figure 4.3). NEP will thus be minimum at the peak wavelength and increase slightly as the wavelength of the photons increase or decrease.

The Receiver (Silicon Photosensor)

The formula for calculating the thermal noise is:

$$\frac{I_{N(Thermal)}}{\sqrt{B}} = 25.3\sqrt{I_s(nA)} \quad (fA/\sqrt{Hz})$$

where

$\frac{I_N}{\sqrt{B}}$ = bandwidth normalized noise current in femtoamps per root hertz

I_s = reverse saturation current in nanoamps

A sample calculation of thermal noise from the stated formula will be useful in the understanding of the magnitude of the noise. A PIN diode with an effective area of 0.065 cm^2 (0.100 in. × 0.100 in.) will be used. Leakage current is assumed to be 25×10^{-9} amps.

$$\frac{I_N}{\sqrt{B}} = 25.3 \sqrt{25 \times 10^{-9}}$$

$$= 25.3 \times 1.58 \times 10^{-4}$$

$$= 400 \times 10^{-5} \text{ femtoamps/root hertz}$$

$$= 4.0 \times 10^{-18} \text{ amps/root hertz}$$

The formula for calculating the shot noise is:

$$\frac{I_{N(shot)}}{\sqrt{B}} = 17.9\sqrt{I_{dc}(nA)} \quad (fA/\sqrt{Hz})$$

I_{dc} = total dark current in nanoamps

$\frac{I_{N(shot)}}{\sqrt{B}}$ = band width normalized noise current in femtoamps per root hertz

A sample calculation of shot noise from the stated formula will be useful in the understanding of the magnitude of the noise. A PIN diode with an affective area of 0.065 cm^2 (0.100 in. × 0.100 in.) will be used. Leakage current is assumed to be 25×10^{-9} amps.

$$\frac{I_N}{\sqrt{B}} = 17.9 \sqrt{25 \times 10^{-9}}$$

$$= 17.9 \times 1.58 \times 10^{-4}$$

$$= 28 \times 10^{-4} \text{ femtoamps/root hertz}$$

$$= 28 \times 10^{-19} \text{ amps/root hertz}$$

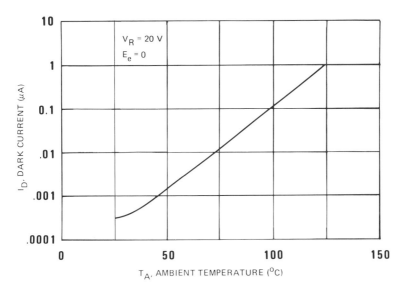

Figure 4.8 Dark current versus ambient temperature. The shape of this curve is similar to a plot of bulk leakage versus temperature. The dark current is composed of a fixed amount of surface linkage plus bulk leakage, which approximately doubles for each 10°C increase in temperature.

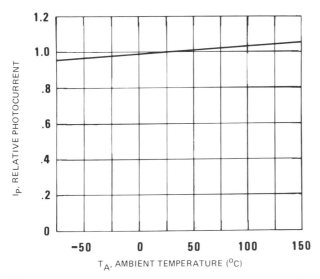

Figure 4.9 Relative photocurrent versus ambient temperature. As with dark current, only less dramatically so, relative photocurrent increases with increasing temperature. In the case of phototransistors, on-state collector current varies with temperature to a much greater extent, and in direct proportion to, the h_{FE} (current gain) of the device.

The formula for calculation of NEP is:

$$NEP = \frac{I_N/\sqrt{B}}{R_\phi} \quad (fW/\sqrt{Hz})$$

where

NEP = the radiant signal flux at a specified wavelength required for unity signal to noise ratio normalized for bandwidth

$\frac{I_N}{\sqrt{B}}$ = bandwidth normalized noise

R_ϕ = flux responsivity in amps/watt at a given wavelength

A sample calculation for NEP from the stated formula will be useful in the understanding of the magnitude of NEP. A PIN diode with an effective area of 0.065 cm^2 (0.100 in. × 0.100 in.) will be used. A flux responsivity of 0.6 amps/W will be used.

$$NEP = \frac{28 \times 10^{-19}}{0.6}$$
$$= 47 \times 10^{-19}$$

The leakage current or dark current (I_{dc}) is the major drawback to operation in the photoconductive mode. This current refers to the current that flows when no radiant flux or photons are applied to the photodiode. Dark current consists of two components:

1. Surface leakage
2. Bulk leakage

The surface leakage of a diode approximates a fixed number while the bulk leakage approximately doubles for each 10°C increase in temperature. When this is plotted on log linear paper, it results in a curve similar to Figure 4.8. At 25°C and below the surface leakage will usually dominate, but at high temperature the bulk leakage will dominate.

The photocurrent (I_p) of the photodiode will vary with temperature. This positive temperature coefficient is shown in Figure 4.9.

This change is a significantly lower percentage than is found in multiple junction photosensors units such as a phototransistor. The leakage in a phototransistor is the diode leakage multiplied by the h_{FE} (current gain) corresponding to that collector currect and bias voltage level.

5

The Phototransistor and Photodarlington

5.1 BASIC THEORY

A phototransistor operates in a manner similar to a conventional small signal transistor except the base or control current can come from both impinging photons on the depletion region or traditional forward bias of the base-emitter junction. The impinging photons utilize the depletion region formed by the reverse biased collector-base junction to create photocurrent, which acts as base current or control current. Most phototransistors are NPN types since this material lends itself to both ease of manufacture and useful electrical parameters. The hole-electron pairs created in the collector-base depletion region cause photocurrent. The electrons move toward the collector or N-type region while the holes move toward the base or P-type region. As described in the basic theory of the P-N junction covered in Section 4.1, Figure 5.1 shows distribution of impurity concentrations in a diffused NPN transistor.

The series of statements made below summarize the basic dimensions and concentration levels within the semiconductor. This information corresponds to the data presented.

Single crystal silicon has a density of 5×10^{22} atoms/cm^3.
Normal N-type emitter impurity concentration at the surface is 5×10^{20} atoms/cm^3 at the surface.
The emitter-base junction is approximately 0.0002 in. deep.
The base width is approximately 0.0002 in.
The collector base junction is approximately 0.0004 in. deep.
Normal P-type base impurity concentration at the surface is 1×10^{19} atoms/cm^3.
Normal N-type starting material impurity concentration is 1×10^{15} atoms/cm^3.
The finished NPN transistor is usually 0.008 in. thick.
The collector region is approximately 0.0076. in. thick.

Figure 5.2 shows the collector current versus collector emitter voltages for varying levels of irradiance (or photons) impinging on the base of the NPN transistor.

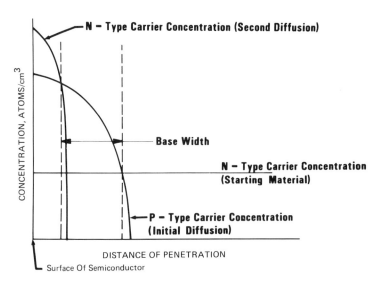

Figure 5.1 Impurity concentration versus penetration for NPN transistor. To form a transistor, P-type inpurities are initially diffused into N-type starting material. N-type material is then diffused into the newly formed P-type material, leaving two junctions, or an NPN transistor.

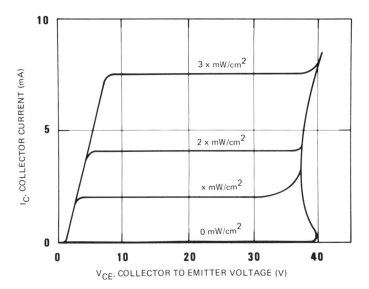

Figure 5.2 Collector to emitter voltage versus collector current. This family of curves is analogous to traditional transistor I_C versus V_{CE} curves except that incident radiation on the base substitutes for base current. Each step in photon level is the analog to an incremental increase in base current, resulting in a family of curves.

5. The Phototransistor and Photodarlington

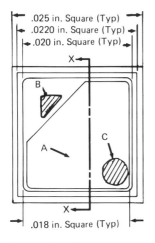
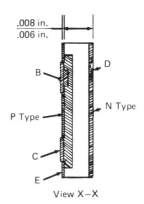

A Active Area (Base)
B Emitter Contact (Al)
C Base Contact (Al)
D Collector Contact (Gold)
E Oxide and Nitride Passivation

Figure 5.3   Top and side views of an NPN phototransistor. In an NPN phototransistor, the maximum top area possible is allotted to base diffusion in order provide as much light sensitive area as possible; however, this compromises some other performance measures, such as frequency response.

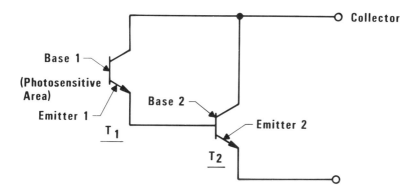

Figure 5.4   Darlington or cascaded transistor schematic. The emitter of the photosensitive transistor drives the base of the second transistor, approximately squaring the h_{FE}. Current gains from 1000 to 25,000 are possible; however, operating speed is almost always traded away to achieve this high gain.

With zero photons impinging on the base region, the collector-to-emitter voltage is gradually increased, and only leakage current (dark current) will flow until breakdown is reached. If the quantity of photons is increased to a certain level and stabilized, and the photon level is then increased again and stabilized with a corresponding increase of collector to emitter voltage, a family of characteristic curves will be formed. The current gain of the transister is the collector current divided by the base current at a particular voltage. The base current corresponds to photocurrent (I_p) discussed Section 4.2 on photocurrent for PN diodes. These curves will be discussed in more detail in the characterization section.

Figure 5.3 shows the top and side view of a phototransistor chip. The emitter diffusion is the triangular shape in the upper left-hand corner of the top view (outlined inside the triangular contact metallization).

The phototransistor is optimized for a large exposed base area to increase the quantity of photons that can strike it. This creates a higher than normal capacitance that compromises the frequency performance of the unit. This will be discussed in Section 5.2.

Figure 5.4 shows the schematic of a Darlington transistor. The unit has a common collector with the emitter of the input transistor driving the base of the output transistor. This effectively squares the current gain or sensitivity but significantly decreases the switching performance. This type of unit is sometimes referred to as "cascaded transistors."

The current gain or h_{FE} of a phototransistor typically can be varied from 50 to 1000 by control of base width. The Darlington phototransistor usually varies from 1000 to 25,000. When gains are needed below 1000 the phototransistor is used, while gains above 25,000 usually create unstable units (particularly at high temperatures). The photodarlington transistor can be constructed with high sensitivity to low light levels by making the base area of the input transistor large. Figure 5.5a is the top view photograph of this type of Darlington transistor. The output transistor emitter and base occupy the upper left-hand corner of the chips. The small indentations on the emitter of the output transistor and the base of the input transistor are electrical probe marks where the chip was electrically tested. Figure 5.5b is the top of a photodarlington transistor that is optimized for higher current operation. The photosensistive base area of the input transistor is the center square with the emitter diffusion into the four corners. The emitter of the output transistor is horseshoe shaped around it. This photodarlington transistor will handle currents in the 100 to 200 mA range while the unit optimized for high sensitivity will work best in the 1 to 10 mA range. Again the small indentations at 6 o'clock and 12 o'clock are electrical probe marks for the base of the input transistor at 6 o'clock and the emitter of the output transistor at 12 o'clock.

5. The Phototransistor and Photodarlington

Figure 5.5 Photodarlington transistors. The large photosensitive area causes these devices to exhibit high sensitivity at low light levels. Note the relatively small size of the output transistor emitter and base, occupying the upper left-hand corner of the chip. The second photograph shows a photodarlington designed for higher current operation, rather than sensitivity. In this case, the output transistor emitter occupies the horseshoe shaped area.

The basic construction method for these types of transistors employs standard bipolar technology. This will be discussed in Chapter 6.

5.2 CHARACTERIZATION

The NPN transistor has two junctions, the collector-base and the base-emitter. The reverse breakdown of the collector-base junction is shown in Figure 5.6.

As the voltage is increased above zero a small leakage current identical to the dark current in the PN diode is observed. This leakage current will increase slightly with increasing voltage. At some point determined by the impurity concentration and the change (gradient) of this impurity concentration versus distance, the depletion region becomes wide enough that a carrier (hole or electron) traveling through it will strike and dislodge a like carrier from the lattice structure of the atom. Increasing the voltage further will cause high currents to flow. Under this condition, the heat generated can cause damage to the device.

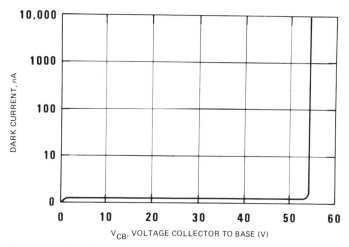

Figure 5.6 Reverse breakdown of collector-base junction. Exactly analogous to diode behavior, reverse breakdown of the collector-base junction is characterized first by the occurrence of a small leakage current as the collector-base voltage rises above zero. The depletion region becomes wider as more voltage is applied, until at breakdown carriers passing through this region will dislodge additional like carriers causing high currents to flow. Any additional increase in voltage at this point would heat and damage the device.

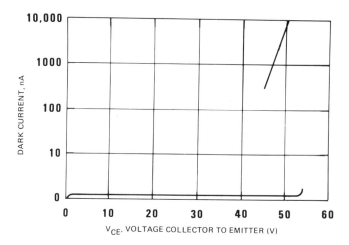

Figure 5.7 Reverse breakdown of collector to emitter junctions. Again, the collector-base junction is reverse biased; however, the emitter-base junction is simultaneously forward biased. At high voltages, the wide collector-base depletion region causes the base to become very narrow, and current gain (h_{FE}) then approaches infinity. New carriers are actually created until the depletion region narrows (the base widens) sufficiently to cause h_{FE} to drop back. V_{CE} also drops back when this stabilization process occurs. Continued high voltages will heat and damage the device.

5. The Phototransistor and Photodarlington

If this voltage is applied collector to emitter, a similar phenomena happens at low voltage. This is shown in Figure 5.7. The observed leakage current is higher due to h_{FE} caused by the forward bias on the base-emitter junction; but since h_{FE} is very low at low currents, the leakage current is only slightly higher than the leakage current observed on the collector base diode.

As the voltage approaches the avalanche voltage (the sum of the voltage drop across the forward biased emitter-base diode, and the voltage drop across the reversed biased collector-base diode), a similar mechanism happens. Now, however, due to the narrow base width created by the wide depletion region at high voltages, when avalanche occurs, alpha goes to one and beta approaches infinity (beta or h_{FE} which is equal to alpha divided by one minus alpha). Carriers are actually being created. Because this is an unstable condition, the depletion region narrows (voltage drop decreases) until alpha drops back to below unity. The collector-base junction is normally specified with a low leakage (~ 100 nA) at the operating voltage and a higher value of leakage (~ 100 μA) to show breakdown (avalanche). The collector emitter junctions are also normally specified with a similarly low leakage (~ 100 nA at operating voltage) but a significantly higher current (~ 1 to 10 mA) to ensure the voltage read is in the stabilized region after alpha has dropped back to below unity.

If now the emitter-base junction is reversed biased, a normal avalanche will occur at low voltage (~ 5 to 7 V). If the reverse bias is applied emitter to collector, the snapback phenomenon will not be seen. The inverse alpha is significantly lower and the depletion region is much more narrow (due to the lower applied voltages).

The leakage current (collector to emitter) constitutes one boundary of the unit when it is operated as a switch. When the transistor is off, the value of the leakage or dark current flowing through the load resistor prevents the unit from being totally "off." This leakage current is usually specified at room temperature. As the temperature is increased, the leakage current will increase. The leakage current of the collector-base junction is multiplied by beta (h_{FE}). Since beta is increasing with both increasing collector current and increasing temperature, the observed leakages are significantly higher than those observed for a PN or PIN diode. Figure 5.8 shows this increase on a log-linear graph.

Increasing the voltage will cause a corresponding increase in leakage current. Refer back to Figure 5.7 for reference.

The phototransistor operated as a switch has two boundary condiditions: the "off" position (controlled by leakage current) and the "on" position (controlled by saturation voltage). Figure 5.9 shows these two boundary conditions when the load is 1000 Ω and the supply voltage is 5 V.

The Receiver (Silicon Photosensor)

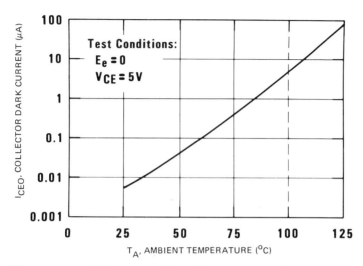

Figure 5.8 Collector-emitter leakage (dark) current versus ambient temperature. When the phototransistor is used as a switch, collector dark current becomes the "off" state or lower limit. Dark current is typically higher than for photodiodes because leakage current is multiplied by the current gain (h_{FE}) of the device.

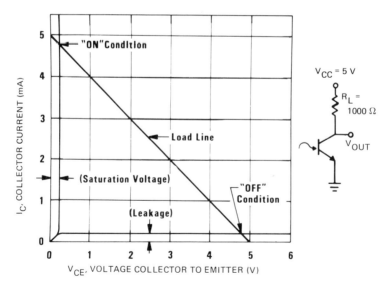

Figure 5.9 Boundary conditions for an optical switch. A "perfect" switch would be "off" at V_{CE} equal to 5.0 volts and "on" at V_{CE} equal to 0 V. Leakage current and saturation voltage limit the device to less than 5.0 V off and more than 0 V on for the circuit shown.

5. The Phototransistor and Photodarlington

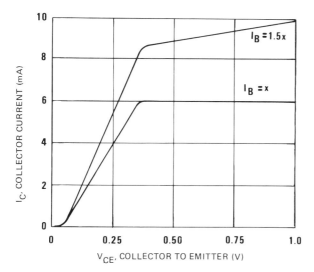

Figure 5.10 Saturation voltage versus collector current. For the $I_B = X$ curve, the initial off-set results from the difference between forward and inverse current gain. The characteristic of the second region is created by device geometry and bulk resistance in the collector. The third portion is the normal linear operating region.

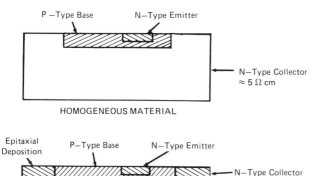

Figure 5.11 Comparison of collector bulk resistance on epitaxial and homogeneous material. In actual practice, the epitaxial deposition is approximately 1.0 mils thick with the total transistor chip is 8.0 mils thick.

The Receiver (Silicon Photosensor)

In a typical example, if the leakage current was 1 μA and the saturation voltage was 0.2 V then the "off" condition would be 4.999 V and the "on" condition would be 0.2 V. The voltage drop across the load would be 0.001 V in the off condition and 4.8 volts in the on condition. The unit would not be a perfect switch (5 V to 0 V) by 1 mV in the "off" condition and 200 mV in the "on" condition. Further analysis of the saturation characteristics is required for understanding this phenomenon. Figure 5.10 shows an exaggerated view of the saturation voltage versus collector current for two different levels of base current.

If we trace the I_b = 1 line from 0 V to 0.75 V, we observe three different regions of saturation voltage. The initial offset voltage is approximately 30 mV and is due to the inverse h_{FE} being lower than the forward h_{FE}. (If inverse and forward h_{FE} were identical then there would be no offset.) The offset voltage is fairly consistent between units made by the same techniques. Since most commercial devices are planar diffused (causing the emitter to be more heavily doped and having a graded junction), a value of 30 to 50 mV is usually considered typical. The second region has a resistive slope controlled by the device geometry and the bulk resistance in the collector region. A common technique for lowering this resistance is to use epitaxial material. Figure 5.11 shows how this significantly lowers the collector bulk resistance.

The third region of saturation voltage is the normal linear operating region. It will be discussed in more detail in the h_{FE} section. The second saturation voltage line of I_B = 1.5x is similar to the I_B = 1x line. Its normal linear operating region is at a higher collector current, which is reasonable if one assumes h_{FE} is relatively constant. The changes in h_{FE} will be discussed later in the chapter. The important point is that the higher the base current for a given collector current, the lower the saturation voltage. This is an important factor since a production spread of h_{FE}'s is normal. Figure 5.12 shows these three regions in a planar epitaxial phototransistor.

The photodarlington will behave in a similar fashion. Since the leakage (dark) current is multiplied by one extra h_{FE}, the value will be larger and will increase more rapidly with temperature. Figure 5.13 shows this increase graphically.

The saturation voltage will also increase since a forward biased diode is added to the saturated circuitry. The offset voltage will increase by approximately 0.6 V and the slope of the saturation voltage will also increase. The net result will be to increase the saturation voltage to approximately 0.63 V at zero collector current and to 0.9 V at 5 mA. These values will vary as the geometry of the devices varies. The saturation resistance has more impact on the total voltage drop at high currents than the forward voltage drop of the base emitter diode.

The current gain or beta has a number of variables that cause it

5. The Phototransistor and Photodarlington

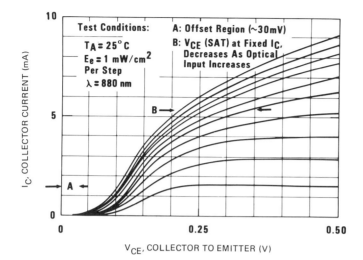

Figure 5.12 Collector current versus collector to emitter voltage. The offset region (A) is largely independent of the irradiance and is primarily a function of the difference between h_{FE} and forward h_{FE}. The irradiance level, as the source of base current, dramatically effects device performance in the operating range (B).

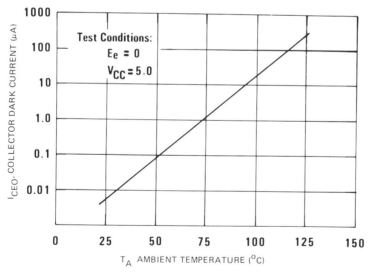

Figure 5.13 Collector dark current (photodarlington) versus ambient temperature. Just as in Figure 5.8, dark current increases with temperature and is then multiplied by the gain of the device. This effect is even more dramatic for the photodarlington (depicted above) due to the additional current gain of the output transistor.

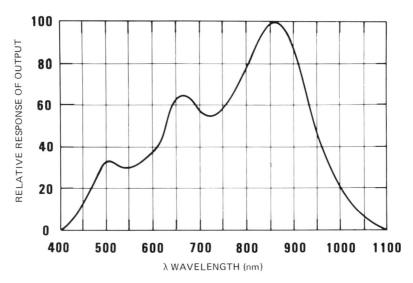

Figure 5.14 Spectral response for NPN phototransistor. An optical overcoating is often applied to help center the peak sensitivity at 880 nm. The two secondary peaks are also created as a result of this overcoating.

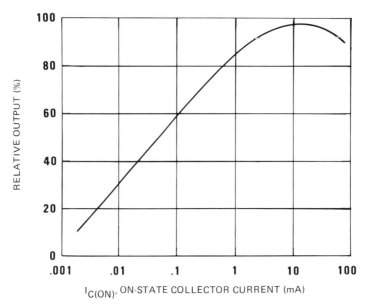

Figure 5.15 Relative output versus collector current. The key to successful use of phototransistors lies in the relationship between changing incident radiation and the on-state collector current. Note that this curve would assume a given collector-emitter voltage.

5. The Phototransistor and Photodarlington

to change. In the section on quantum efficiency of the PN diode (Section 4.2), it was shown that for a PIN diode with a peak response at 900 nm, the maximum quantum efficiency (electrons per photon) was 83%. As the wavelength increased or decreased the quantum efficiency decreased. Since the phototransistor has a narrower depletion region located closer to the surface, the peak response will be both lower and less efficient. This response versus wavelength is shown in Figure 5.14. The response is modified by the addition of an antireflective coating deposited on top of the phososensitive area and designed to enhance the IR response. This quarter wave overcoat is designed to maximize device sensitivity for wavelengths of 880 nm. The two local maxima to the left of the 880 nm global maximum are also the result of the optical overcoating and are analogous to the secondary resonance peaks found in acoustics. Although theory predicts that these secondary peaks should appear at 440 nm (quarter wave) and 660 nm wavelengths, in reality they are confirmed at 475 nm and 660 nm.

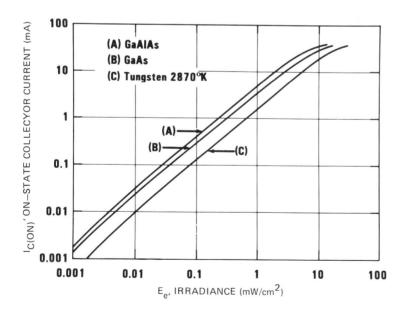

Figure 5.16 On-state collector current versus irradiance. Phototransistor collector current is typically a linear function of irradiance in the range shown. Design differences (e.g., lens changes) will shift the location of these curves but not their linear characteristics.

The peak spectral response is effectively lowered as the diffused junctions move closer to the surface. This is done in order to improve the frequency response of the phototransistor. This further lowers the quantum efficiency. The normal topographical layout of a phototransistor optimizes the exposure of base material to improve responsivity. The resulting phototransistor has a reasonably linear response in the 500 μA to 20 mA collector current range. This is normally adequate since smaller geometries would not improve the response at the lower end and the photon-h_{FE} mix would not make the maximum end any more usable. The lower light levels are usually detected by a photodiode-amplifier combination or in some cases by a photodarlington. The upper end of output current requirements can be extended by a photodarlington with a high power output transistor or by feeding the output of the phototransistor into another silicon transistor. Figure 5.15 shows relative IRED output versus collector current for a typical phototransistor.

Different energy sources with their different spectral content cause a phototransistor to have different outputs. Figure 5.16 shows the $I_{C(ON)}$ of a narrow included acceptance angle phototransistor (25° between half power points) encapsulated in a plastic TO-18 outline package. This graph will shift with different sensitivity levels of photosensors and variations in energy sources among different manufacturers, but the shape remain relatively consistent.

The phototransistor collector current has a positive temperature coefficient. As the h_{FE} decreases, this curve will tend to flatten, and as it increases, it will become steeper. When the h_{FE} is in the 400 to 600 range, then the curve will nearly cancel the IRED output-temperature curve so the resultant or coupled output is nearly flat. Since the summation of these curves approximates a multiplication rather than an addition, the curve will tail down at low temperature (transistor dominated) and tail down again at high temperature (IRED dominated). This is the curve most often shown on manufacturers' data sheets (refer to Figure 10.8). Figure 5.17 shows this change for a unit with h_{FE} in the 400 to 600 range.

The switching time of a phototransistor is significantly longer than its small signal transistor equivalent with equal current carrying capability. The major reason is the significant increase in the capacitance of the collector-base diode. In a phototransistor, the area of the base exposed to impinging photons is greatly increased in order to enlarge the energy gathering area and improve device sensitivity. Figure 5.18 shows the comparable sizes for a phototransistor and its equivalent current-carrying small signal transistor.

The base area on the phototransistor is 484 sq. mils. while the base area on the equivalent transistor is only 132 sq. mils, a factor of 3.67 larger. With equivalent diffusions, this ratio would be an accurate predictor of the difference in capacitance. In actual prac-

5. The Phototransistor and Photodarlington

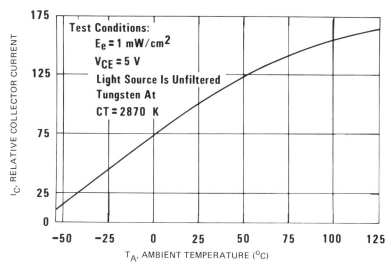

Figure 5.17 Normalized collector current versus ambient temperature. With a constant source of infrared energy, higher operating temperature results in higher relative output current. The opposite behavior occurs for relative output versus temperature for IREDs making possible coupled configurations with lower overall temperature drifting.

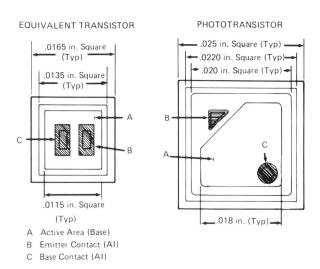

Figure 5.18 An NPN phototransistor and its equivalent NPN small signal transistor. The base area (A) is 3.67 times larger for the phototransistor shown. While this allows for a much greater light gathering area, base to collector capacitance is increased resulting in slower speed than conventional transistors.

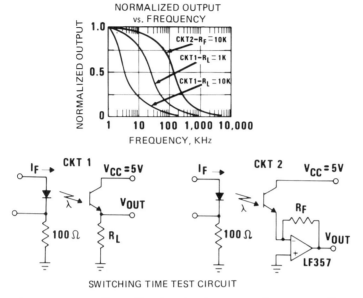

Figure 5.19 Normalized output versus frequency for an NPN phototransistor. Starting from the left, the lower load resistance (by an order of magnitude) results in a tenfold frequency response increase. The use of a linear amplifier in test circuit two affords the fastest frequency response.

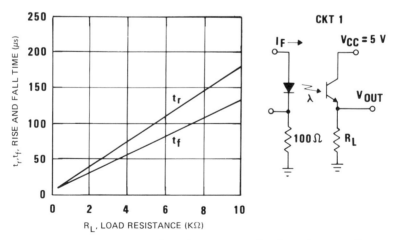

Figure 5.20 Rise and fall time versus load resistance for an NPN phototransistor. A model depicting charging or discharging a capacitor through a resistor (RC network) would predict faster speed at lower resistance, just as shown in the relationship between load resistance and rise/fall times or frequency response.

5. The Phototransistor and Photodarlington

tice, tighter masking tolerance is used on the transistor and the ratio is typically even higher. In addition, the diffusion profile on the phototransistor is deeper in order to improve the photon wavelength response in the infrared region. This further aggravates the problem of bandwidth response for the phototransistor. Figure 5.19a, b and Figure 5.20 show the results of this higher capacitance and resulting lower bandwidth response.

The switching time of a photodarlington transistor is significantly slower than the transistor counterpart. Figure 5.21 shows the switching time versus load resistance. Note that the times are given in milleseconds rather than microseconds.

The lens characteristics of the phototransistors are similar to their LED counterparts. The photosensitive area of the chip is generally larger than the emitting size of the IRED, but the phototransistor normally does not set in a well. A typical well diameter is 0.030 in. while a typical photosensitive area diameter is 0.022 in. The net result is the effective size of the emitting area of an IRED is close to the receiving area of the phototransistor. The subtle differences, however, lead to physical differences in the outline of the two packages. Figure 5.22 shows receiving angle pattern plots and side view drawings of a wide beam angle plastic. Figure 5.23 shows comparable drawings for the metal package.

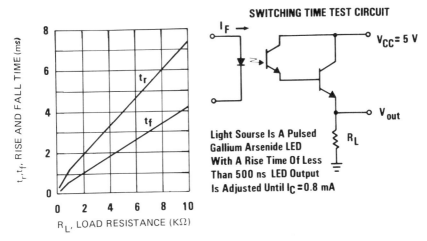

Figure 5.21 Switching time versus load resistance for a photodarlington transistor. More than an order of magnitude in switching speed over phototransistors is traded for higher output currents in a typical photodarlington; however, in simple motion sensing applications where speed is not critical, the component savings may be well worth the tradeoff.

The Receiver (Silicon Photosensor)

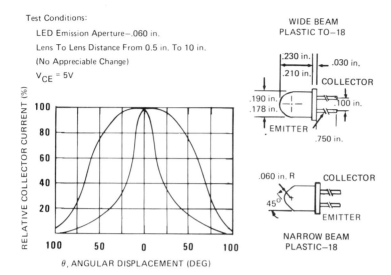

Figure 5.22 Receiving pattern plots and side view drawings of wide and narrow beam plastic TO-18 outline packages. These simply represent the inverse of the IRED beam pattern plots shown earlier. To create these patterns, a collimated source is rotated through an arc in front of the device while $I_{C(ON)}$ is measured.

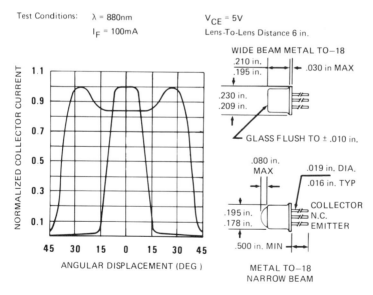

Figure 5.23 Beam pattern plots and side view drawings of metal TO-18 outline package. The wide angle "doughnut" effect results from reflections off the side of the metal can.

6
The Photointegrated Circuit

6.1 BASIC THEORY

Chapter 3 showed that the PN and PIN diodes were significantly faster in speed than the basic phototransistor. A review of Chapter 2 leads one to the conclusion that the phototransistor adds gain to the photodiode but compromises the speed performance and the photodarlington advances the trade-off even further. In many applications, the speed-gain relationship of the phototransistor is the most cost-effective method of solving the system design problem. The photointegrated circuit (photo IC) offers two other dimensions of control that can be very useful in certain applications. In many designs, the photoswitch must be in a remote location from the processing electronics. In order to achieve precise switching control (resolution) the photosensitive area may also be reduced. This type of application will be discussed in more detail in Part III. The resultant small electrical signal could cause reliability problems. If the signal path back to the processing electronics is long and there is a high probability of erroneous signals caused by coupled noise from motors, relays, or other forms of electrical interference, the design will require a higher signal level than is possible from the apertured phototransistor. The photo IC then becomes the most cost effective solution to this type of problem.

Rotary or linear encoders of high resolution are gradually shifting toward use of the photo IC to take advantage of the inherent speed, the reduced system space, and the cost reduction attained over systems fabricated from discrete devices. At the present time, *bipolar* technology is widely used to fabricate photo ICs. Figure 6.1 shows a simple photo IC. This circuit will be used for discussion showing the basic construction used in bipolar technology.

The photodiode is made as a PN junction (similar to the collector-base diode) with the P-type material exposed for photon coupling. The N-type region is electrically common to the collectors of the two NPN amplifying transistors. The bypass resistor must be electrically isolated from the other elements for the circuit to function properly. Figure 6.2 shows the top view layout of the photo IC.

The Receiver (Silicon Photosensor)

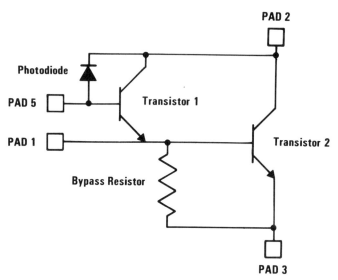

Figure 6.1 Schematic of photo IC with transistors, resistor, and photodiode. For the purpose of illustration, the IC fabrication process, a simple photo IC circuit is presented.

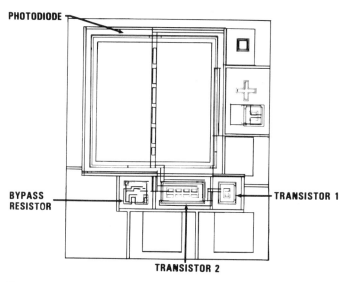

Figure 6.2 Top view layout of photo I.C. Even a simple circuit quickly begins to look complex when viewing its layout. Identifying the individual components without knowledge of the circuit diagram is a difficult job even for the trained eye.

6. The Photointegrated Circuit

Comparing Figures 6.1 and 6.2 is difficult unless the reader is familar with bipolar processing technology. In order to bridge the gap between the two figures, it is necessary to review the construction of the IC itself. The approach used will be to follow a simplified overview of the various processes utilized in the fabrication of the finished chip. The construction of the output transistor (transistor 2) will be used as a focus for the various steps. First, the starting slice of P-type silicon is oxidized. This is done by placing it in a furnace with a surplus of oxygen molecules. A layer of SiO_2 is grown on the slice. A thin layer of photographic emulsion is then placed over the entire slice surface. Selected areas of the photographic emulsion are exposed to ultraviolet light. This is virtually identical to the process used in photography. The unexposed photoresist (photographic emulsion) is then removed. The silicon dioxide, in areas where no photoresist is located, is chemically removed. Figure 6.3 shows a cross section of the silicon slice depicting this process sequence.

After the exposed photoresist has been removed, the slice is then placed in another furnace and coated with a layer of N-type impurities. At high temperature, these N-type impurities diffuse into the surface. This process is similar in principle to a drop of ink dropped into a glass of still water. The ink drop (high concentration) will gradually spread or diffuse into the surrounding water. The diffusion rate will double for each 10°C rise in temperature. Of key importance is the fact that the diffusion rate into the oxide is much lower or slower than the diffusion rate into the silicon. If the oxide was grown thick enough, it will act as a diffusion barrier giving selective penetration of the N-type impurity atoms. The localized N^+-type areas will become the areas located beneath the active components.

They will be used to lower the series resistance in the N-type region. Figure 6.4 shows the top and side views of these areas as a cross-hatched region. The oxide areas have been removed.

The N^+ type nomenclature is used to show a heavier than normal concentration of N-type impurities. It is comparable to an emitter diffusion of low resistance per unit area. The slice is then placed in an epitaxial reactor and an N-type region approximately 0.0005" thick is grown on the slice. The process is similar in function to the epitaxial layer grown on the GaAs substrate for IRED fabrication except the method uses gaseous phase rather than solution phase material. Figure 6.5 shows the same cross-sectional side view with the addition of the N-type epitaxial layer. The slice thickness is approximately 0.014 in. and the N^+-type material is approximately 0.0002 in.

Another coat of oxide is grown over the entire slice. The side with epitaxial N^+-type material will be the only side considered. An oxide removal is performed selectively removing the oxide in a narrow

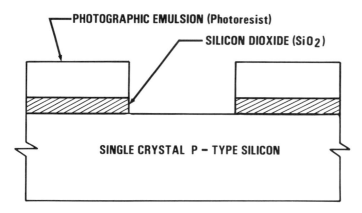

Figure 6.3 Silicon slice after an oxide removal (OR) step. Photolithographic techniques allow for precise placement of oxide coverings. The wafer is now ready for diffusion of N-type impurities into areas slated to become starting material for active components.

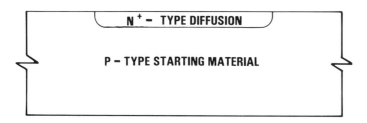

Figure 6.4 Photo I.C. transistor after completion of N-type diffusion and oxide removal. The N^+-type material has a heavier concentration of N-type impurities than normal N-type material. This creates low resistance per unit area, and as such, these regions will lower the series resistance for the IC.

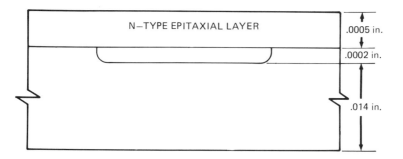

Figure 6.5 Photo IC transistor after epitaxial layer growth. In a manner similar to IRED epitaxial growth, a 0.0005 in. layer of N-type material is grown on the wafer. The N^+-type region now becomes known as the "buried layer" (drawing not to scale).

6. The Photointegrated Circuit

region above the edges of the "buried layer" N^+-type diffusion. Another N^+-type diffusion is performed. The net effect is to create a low resistance path for the carriers traveling through the N-type anode or collector regions back to the top surface contact. Figure 6.6 shows the top view of the added diffusion only while the side view shows a cutaway view portraying the original buried layer, the added epitaxial growth, and the deep N^+-type diffusion.

Another oxide removal is performed that selectively removes the oxide in a narrow region bounding the active components, the resistor, and the bond pads. It opens up the area between adjacent circuits. This is to allow ease of processing in subsequent circuit separation when the individual chips are segregated. A P^+-type diffusion is performed. This creates N-type tubs that are electrically isolated from each other by a reverse biased diode.

Figure 6.7 shows the top view of the added diffusion only, while the side view shows a cutaway view that portrays all processing steps to this point.

Another oxide removal step is performed that selectively removes the oxide in a large area. This area is slightly smaller than the isolated islands of N-type material. A P-type diffusion is made through this opening. This will become the base region of the transistors and the exposed photosensitive region of the input photodiode. Figure 6.8 shows a cutaway view that portrays all processing steps up to this point.

Another oxide removal is performed that selectively removes the oxide in four small areas. These will become the emitters of parallel NPN transistors. The transistor requires both low saturation voltage and good current carrying capability in the "on" condition. The technique of effectively paralleling a number of transistors (in this case four) is widely used to enhance the current carrying capability. Figure 6.9 shows the side view. The new N^+-type diffusion is cross hatched for clarity.

The NPN transistor is now structually complete. Figure 6.9 has been redrawn to detail the P-type and N-type areas. The N-type and N^+-type areas are cross-hatched. The P^+-type and P-type areas are left clear. Figure 6.10 shows this in both the top and side views.

Another oxide removal is performed that selectively removes the oxide from all the areas where electrical contact is desired. The contact metallization is then deposited and the undesired metal removed in a subsequent photoresist masking process. The metal may be deposited by evaporation, sputtering, or electron beam deposition. The metals utilized may actually consist of more than one layer of metals but is usually a single layer of aluminum. The deposition systems and the type of conductors will vary widely from one manufacturer to another. Figure 6.11 shows the top view of the output transistor. The base contact egresses to the left, while the emitter collector contacts egress to the right.

The Receiver (Silicon Photosensor)

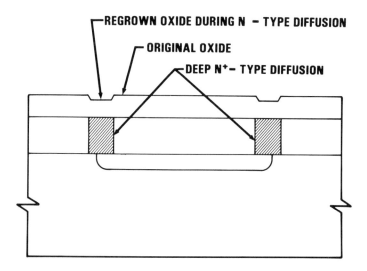

Figure 6.6 Photo IC transistor after deep N^+-type diffusion. The photolithographic oxide process described earlier is repeated and deep N^+-type diffusion regions are added. These result in low resistance regions to connect the surface to the active regions of the buried layer.

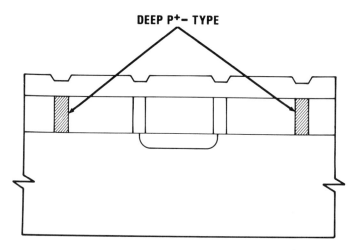

Figure 6.7 Photo IC transistor after deep P^+-type diffusion. An N-type "tub" lies inside the P^+-type deep diffusions. The P^+-type areas will act as reverse bias diodes to electrically isolate the various active areas.

6. The Photointegrated Circuit   105

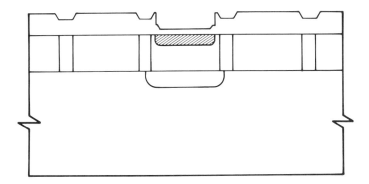

Figure 6.8   Photo IC transistor after P-type base diffusion. The IC is now prepared for addition of N-type areas within the P-type base area. These will form NPN transistors.

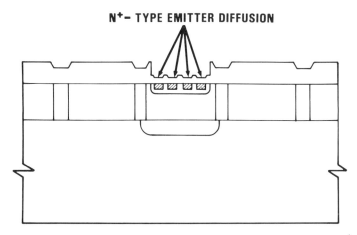

Figure 6.9   Photo IC transistor after completion of N^+-type emitter diffusion. Four N^+-type emitter regions added within the P^+-type base to complete fabrication of parallel NPN transistors. The technique of using four separate emitters enhances the current carrying capability of the IC.

The Receiver (Silicon Photosensor)

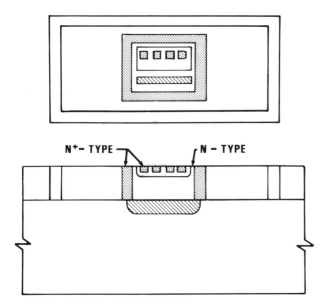

Figure 6.10 Photo IC transistor after completion of N^+-type emitter diffusion with N and P regions highlighted. Figure 6.9 has been redrawn to detail the different regions. No cross-hatch appears in the deep P^+-type "guard-ring" or the P^+-type base area.

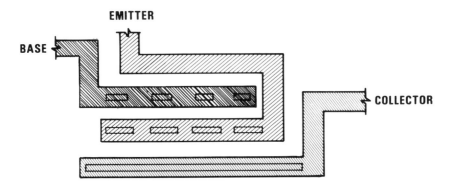

Figure 6.11 Photo IC transistor metallization layout. Metallization techniques vary among manufacturers; however, a single layer of aluminum is most commonly used. Interconnection with collector, base, and emitter is possible with the simple pattern shown.

6. The Photointegrated Circuit

The resistor described in Figure 6.1 may be formed by one or more techniques. If the P-type base diffusion is used, then the oxide removal prior to base diffusion may be used to remove a long thin channel of oxide. The base diffusion sheet resistance is normally near 200 ohms per square (0.001 in. × 0.001 in. area). A 1000 Ω resistor would be 5 squares long (5 mils long × 1 mils wide, or 2.5 mils long × 0.5 mils wide). This technique is normally used for wide tolerance (±10 to 20%), relatively low value (<20K Ω) resistors. The tolerance may be improved by taps or shorting bars across the ends of the resistor. Figure 6.12 shows a 10 KΩ resistor with a ±10% tap.

A second technique utilizes ion implantation of P-type impurities deposited in a thin layer on the surface of the silicon. This technique is similar to the diffused resistors except that the deposited layers are higher in resistivity (up to 2000 Ω/square) and shallower in depth, and better control of resistor value is usually possible. In the circuit shown the resistor is designed for 50K ohms. At 2000

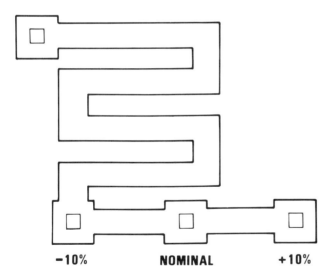

Figure 6.12 A 10 KΩ diffused resistor with ±10% taps. The use of three taps in the example shown allows trimming of the resistor at a probe station. This is done either through use of a current surge to burn out the unwanted section or by selective bonding to the desired tap. In cases where the tap section is likely to be consistent throughout the wafer, selective metallization offers a third method of tap selection.

ohms per square the resistor is 25 squares long. A third technique utilizes thin film deposited resistors. This technique is also similar to the ion implantation method except the deposited metal remains on top of the oxide and the excess deposited material is removed by selective etching. Manufacturers use this technique to lessen the effect of temperature change, which causes the resistor to change in value.

The PN diode is similar in structure to the PN transistor with two major exceptions. It is significantly larger in order to absorb more photons and the emitter N^+-type diffusion is eliminated. The diode or PN junction utilizes the P-type base diffusion as the cathode and the N-type epitaxial layer as the anode. A PNP transistor may be formed by utilizing the N-type epitaxial region as the base and diffusing two P-type regions as parallel bars into the N-type epitaxial region. The N-type base is between these two diffused P-type emitters. The h_{FE} for this is usually low. This structure is shown in Figure 6.13 in both the top and side views.

Finally, if necessary a capacitor may be formed by utilizing the N^+-type emitter diffusion as one plate, a thin deposited oxide as the

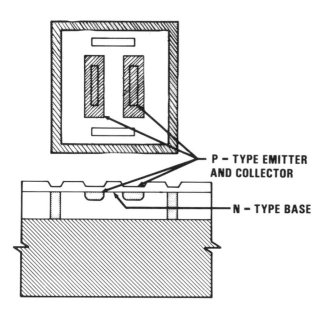

Figure 6.13 Lateral PNP transistor. The N-type material acts as base region between two P-type diffusions acting as collector and emitter. While the transistor is of the PNP configuration, the h_{FE} is usually low.

6. The Photointegrated Circuit

dielectric, and a deposited metal overcoat as the other plate. This is called an MOS (metal-oxide-silicon) capacitor.

Figure 6.2, now appears again, repeated showing the completed IC. The P-type silicon slice has gone through these operations:

N^+-type buried layer diffusion
N-type epitaxial layer deposition
N^+-type deep diffusion
P^+-type isolation diffusion
P-type base diffusion
N^+-type emitter diffusion
P^+-type ion implantation

The oxide over the PN photodiode is a deposited oxide, the thickness of which is closely controlled. On the PN photodiode the oxide acts as a filter for GaAlAs energy, which peaks at 875 nm. The peaks were discussed in Chapter 5 and visually portrayed in Figure 5.13.

Figures 6.3 through 6.13 show how the oxide thickness varies over the entire surface of the photo IC. During the diffusion process additional oxide is regrown. Since each subsequent diffusion is done at a lower temperature to allow minimum additional diffusion depth of the previous diffusions, the grown oxide becomes progressively thinner. The color of the oxide is a function of its thickness. When viewing a photo IC under magnification, the diffused area are seen as different colors. This simply means the different color oxides over the diffused area are the result of thicknesses.

6.2 CHARACTERIZATION

A number of different simple photo IC circuits are discussed below. These are similar in function to the photo transistor and photo-darlington discussed in Chapter 5. As a result of comparison of gain versus speed shows the relative advantage of the photo IC versus phototransistors. Figure 6.14 shows the schematic for a phototransistor and its equivalent photo IC equivalent where the gain of each device is considered equal.

When the load resistance is 1000 Ω the phototransistor will reach its half power point at 25 kHz while the photo IC will attain speeds of 500 kHz. The addition of a second gain stage to the photo IC would lower the frequency response of the photo IC to 25 kHz or equal to the phototransistor. The gain, however, would be increased by two orders of magnitude. These changes are summarized in Table 6.1.

The current carrying capabilities of the devices are equal to those of a photosensitive diode occupying approximately 400 sq. mils. The chip sizes for all three units are similar. This table clearly shows the advantage of the photo IC. This advantage becomes minor when the device is used as a two terminal part where the photodiode

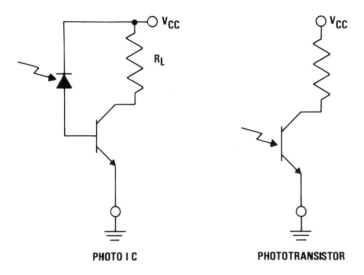

Figure 6.14 Phototransistor and equivalent photo IC. The operating speed of this simple photo IC is over an order of magnitude faster than the phototransistor, and the equivalent gain is possible.

becomes common to the collector of the transistor. The capacitance of the photodiode-transistor would then (equivalent circuitwise) approach the capacitance of the phototransistor negating the speed advantage.

A variety of integrated circuits have been developed that use the output of a photodiode as their input signal. Figure 6.15 shows a high speed (75 ns propagation delay time to high or low output level) photo IC utilized in optocouplers. The photocurrent from

Table 6.1 Relative Gain and Frequency Response of Photo IC versus Phototransistor (1 KΩ load)

	Gain	Half power frequency
Phototransistor	1	25 kHz
Photo IC (single gain stage)	1	500 kHz
Photo IC (double gain stage)	100	25 kHz

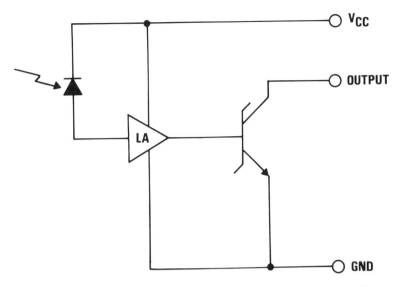

Figure 6.15 A 5 MHz photo IC. More complex photo ICs are possible, offering further improvements in speed and gain. The circuit shown is typical of those used in optoisolators for telecommunications and data transmission applications.

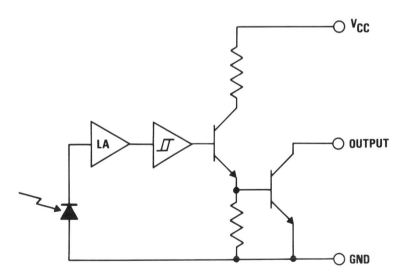

Figure 6.16 A photo IC with hysteresis characteristics. The addition of a Schmitt trigger improves transmission quality and speed for logic applications.

the photodiode is amplified by a high gain linear amplifier, which drives a Schottky clamped open collector output transistor. The unit will sink 13 mA.

Another popular IC incorporates a Schmitt trigger, which provides a hysteresis characteristic to improve the performance with varying input signals. Once the input energy level reaches a certain point the circuit will switch "on." The input energy level must drop significantly below this level for the circuit to switch "off," The ratio of turn "on" to turn "off" is called the hysteresis ratio.

Figure 6.16 shows this circuit. This circuit is shown in the open collector output configuration but is available in the totem pole output configuration. Both types of output are available as buffers or inverters.

The photo IC with hysteresis characteristics is also available with a built-in voltage regulator. This allows the photo IC to be used with supply voltages up to 16 V. Units are available that will go to higher voltages but power dissipation may become excessive. Figure 6.17 shows a commercially available unit in the open collector configuration.

The units shown in Figures 6.16 and 6.17 effectively allow amplication and a hysteresis characteristic internal to the sensor. They are widely used in limit switches and encoders. The maximum speeds are in the 100 to 300 kHz range.

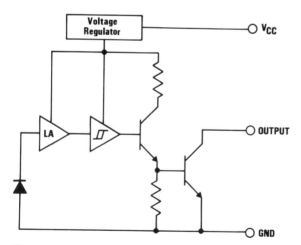

Figure 6.17 A photo IC with internal voltage regulator. The addition of a voltage regulator increases device flexibility and lowers cost by eliminating external supply components. Applications with a remote sensor location such as a motor encoder or computer "mouse" (track balls) are now practical due to this and other photo IC design improvements.

7
Special Function Photo ICs

7.1 BASIC THEORY

Chapter 6 is devoted to a basic understanding of photo ICs and bipolar technology. The relatively simple circuits offer cost-effective solutions to many applications. The photo IC requires more complex processing for chip fabrication and packaging; therefore, the photo IC is more expensive. As volume usage grows, the cost will decrease, but for comparable volumes the photo IC will always demand a premium price over simple phototransistors. The use of photo ICs is increasing due to two strong technological pressures in the marketplace.

The first pressure comes from the user's desire to purchase a complete building block or function. This building block can be pretested by the supplier to a set of specifications and thus reduce the complexity and cost of the final system. A single supplier would control those variables within the specialty skills brought to market by that supplier. In order to illustrate this point, consider a simple switch function. The application requires a flag mounted on a moving object such as a cash register door. The flag would intercept an energy beam from an IRED to photosensor. If the user purchases a discrete IRED, a discrete photosensor, and a molded or machined housing to hold the discretes, then control of at least three suppliers is required. The user must have technical capability both to specify and troubleshoot the units required by the system. This technical capability would be significantly reduced if the complete switch assembly were purchased from a single supplier. The average application is growing in electronic complexity and/or mechanical complexity and volume. As a result, the user is purchasing a more complex part. Photo ICs lead to both improved system performance and reduced system cost.

The second pressure comes from the need for improved performance. The photo IC offers the option to integrate (lower cost/smaller volume) a discrete function or perform a function that is extremely difficult utilizing discrete components. A more thorough analysis of system needs and cost is required to justify this approach in a specific case since a custom photo IC with unique packaging requirements requires a major expenditure in both tooling costs and technical effort. If the photo IC is commercially available, however, the cost comparison becomes very straightforward. Three of these standard function photo ICs will be discussed.

7.2 CHARACTERIZATION

Three special function photo ICs are discussed with respect to application, circuit block diagram, and how they function. An automatic brightness control circuit, a zero current crossing triac driver, and a zero voltage crossing triac driver are covered.

The automatic brightness control uses pulse width modulation to control the brightness of a display system. This simply means that if the display is operating at a 200 Hz rate, a period would be defined as 1/200 sec or 5 ms. The ABC sensor would control the percentage of the 5 ms that the display would be "on." If the initial threshold were set for 2.5 ms, then as the ambient light level to the photo diode increased, the 2.5 ms "on" time would be increased a proportional amount. As the ambient light level decreases the "on" time is proportionally decreased. Since the human eye cannot perceive the "on" or "off" changes as they occur, only a dimming or brightening effect is seen. The end result is that the human eye perceives the display as having the same level of readability independent of the background lighting changes.

Figure 7.1 shows a block diagram of the circuit contained on the monolithic photo IC. It contains an 1834 square mil photodiode, a high gain temperature compensated current amplifier, three comparators, assorted logic functions, an output driver, and a voltage regulator. The unit is mounted in an eight pin clear plastic dual-in-line package and can be used with power supplied ranging from 4.5 to 16 volts.

Table 7.1 gives a listing of the pin functions.

The basic circuit function is discussed in more detail in the following paragraphs. The sensor controls "display brightness" in the following manner: A ramp frequency, determined by an external resistor and capacitor, is generated at pin 5. The output of the ramp generator is internally connected to the "signal" comparator (−input). A control voltage, proportional to received ambient light, is also internally connected to the signal comparator (+input). When the control voltage is greater than the ramp voltage, the comparator is switched on and the output (pin 7) is at a logic "1." These waveforms are shown in Figure 7.2.

The comparator output is squared "up" by going through additional logic gates, which also serve to make sure the sensor is synchronized with an external sink signal, if used.

7. Special Function Photo ICs

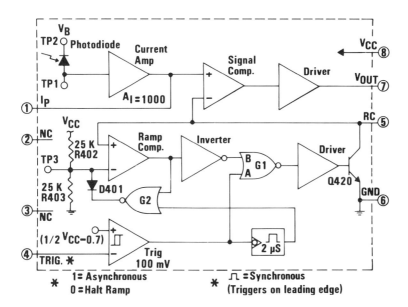

Figure 7.1 Block diagram of an automatic brightness control photo IC. Used to maintain panel display "readability" regardless of the ambient light level, the ABC circuit illustrates the building block philosophy behind photo IC design. The equivalent function would require integrating multiple components from a variety of suppliers at a much higher cost.

It can be seen from Figure 7.2 that the output pulse and the ramp pulse start together. The output pulse continues until the ramp pulse contacts the superimposed line representing the control voltage, V_C. It can also be seen that as ambient light increases, V_C will rise, causing a corresponding increase in pulse width that increases the brightness of the display. The approximate ramp frequently can be determined by using the formula:

$$f = \frac{1.4}{RC}$$

To obtain an asychronous ramp signal, pin 4, TRIG, is connected to pin 8, V_{CC}. This lets the ramp frequency run free with the ramp voltage climbing to $1/2\ V_{CC}$ and then dropping to 0.9 V. To obtain a synchronous ramp signal, apply a sink pulse to pin 4 (see Figure 7.3).

Table 7.1 Listing of Pin Functions for ABC Sensor

Pin 1
Amplified photocurrent is sourced out of pin 1 from an open-collector PNP current source. I_p is 1 to 2 mA under normal room light conditions. In addition to the dimming function, shorting pin 1 to V_{CC}-2V turns the display(s) full "on." Shorting pin 1 to ground forces pin 7 to logic corresponding to a full "off" conditon.

Pin 2
No connection.

Pin 3
No connection.

Pin 4
Tying pin 4 to pin 8 or V_{CC} causes the sawtooth to free-run (asyc-chronous mode). Sawtooth repetition rate is set by the external RC connected to pin 5.

Where synchronous operation is required (e.g., multiplexed digits), its), a pulse is fed into pin 4. The rising edge of the pulse should coincide with the beginning of each digit's "enable" time. This edge causes the sawtooth to stop rising, discharge, then begin its ramp-up.

If pin 4 is continuously grounded, the sawtooth will stop. Thus causes pin 5 to charge to V_{CC} until a change occurs on the trigger input.

Pin 5
This is the point for the timing components that set the sawtooth repetition rate. Typical is 4 µs for a 0.1 µF capacitor C_X. These components can be selected by:

$$f = \frac{1.44}{R_x C_x}$$

Maximum recommended value for R_x is 100k.

Pin 6
Device ground.

Pin 7
The output driver will directly drive the grid of a vacuum fluorescent display. It will source 50 mA at a minimum voltage of V_{CC} = 3 V and will sink 20 mA with a maximum voltage of 0.4 V. Interface to LEDs and incandescent bulbs requires additional components.

Pin 8
Supply voltage operating range is 4.5 to 16 volts. Supply current is typically 6 mA and is relatively constant over this voltage range, with E_e = 0. Supply current will increase with increasing light levels to a maximim of 15 mA.

7. Special Function Photo ICs 117

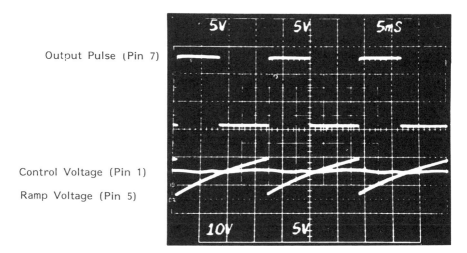

Output Pulse (Pin 7)

Control Voltage (Pin 1)
Ramp Voltage (Pin 5)

Figure 7.2 Waveforms showing how ABC sensors function. Ramp frequency is controlled by an external timing capacitor. The basic control voltage is proportional to the ambient light level detected by a photodiode. It serves as reference for a comparator which, relative to the ramp frequency, determines the percentage "on-time" for the display.

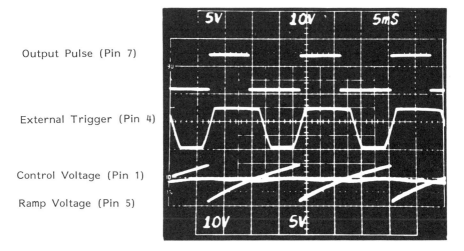

Output Pulse (Pin 7)

External Trigger (Pin 4)

Control Voltage (Pin 1)
Ramp Voltage (Pin 5)

Figure 7.3 Waveforms showing a synchronous ramp signal. In this case, a sync pulse is connected to pin 4 (TRIG) to control certain types of displays where synchronized operation is necessary.

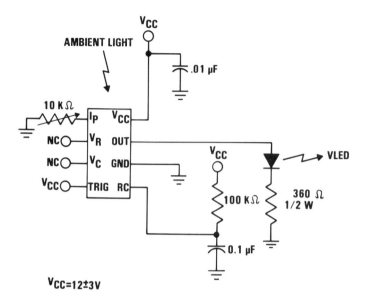

Figure 7.4 Controlling brightness of a VLED. This demonstrates the use of the ABC sensor to adjust the light output of a standard VLED. With a 12 VDC supply voltage, the maximum average current through the VLED would be 32 mA.

The ramp is then synchronized and triggers on the positive going edge of the trigger pulse. The trigger pulse must be greater than $1/2\ V_{CC} - 0.7$ V. Ideally, the asynchronous ramp frequency, determined by the RC constant at pin 5, should be slightly lower than the frequently with which the designer is trying to synchronize. If it is not, the trigger pulse will keep the ramp voltage from climbing to $1/2\ V_{CC}$ and therefore limit the adjustment range of the device. If the asynchronous ramp frequency is set faster than the synchronous frequency and pin 4 is left high after a trigger pulse is received, the sensor will generate an early trigger pulse. This will start the ramp, and it will be restarted when the next trigger pulse is applied. Early triggering is inhibited by logic gate G1 as long as the trigger input is low when the ramp voltage passes through $1/2\ V_{CC}$. The ramp frequency will not vary with supply voltage changes in either direction.

7. Special Function Photo ICs

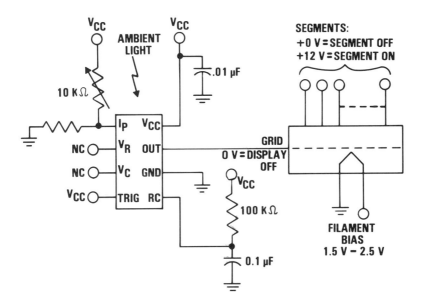

Figure 7.5 This demonstrates the use of the ABC sensor to control a vacuum fluorescent display. The sensor is capable of adjusting the display brightness from 0 to 100%. The sensor should be mounted so that it receives light coming from the direction from which the display will be viewed. Resistor R_1 (tied to I_p) is adjusted for a brightness level pleasing to the eye. The sensor will now adjust for changes in ambient light and make the display appear to remain at the same brightness level.

The ABC sensor is equipped with an externally adjustable light sensitivity level. The sensitivity is set by placing a resistor between pin 1, I_p, and ground. This resistor is the load resistance of the current amplifier. The resistance needed will range from 5 kΩ to 50 kΩ, depending on the ambient light level with which the designer is working.

The ABC sensor is capable of working with power supply voltages ranging from 4.5 to 16 volts. If the sensor is used in a battery controlled system and the battery voltage drops, the sensor will increase its duty cycle and cause the display it is controlling to maintain a constant light output. The sensor is equipped with totem pole output which will sink 20 mA or source up to 50 mA. Care must be taken not to exceed the maximum power dissipation for the particular packaged style being used.

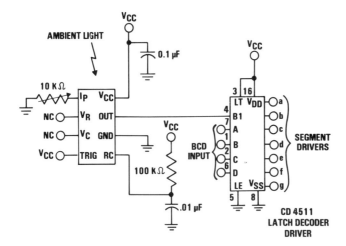

Figure 7.6 This is an example of the ABC sensor controlling a latch decoder driver. If several drivers are used and multiplexed, they can all be controlled by a single ABC sensor. The "strobe" or "enable" line for each display should be connected to pin 4 via appropriate logic circuitry.

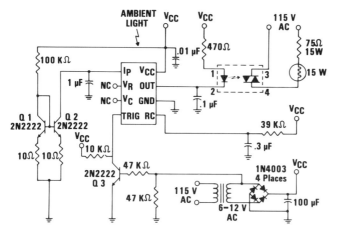

Figure 7.7 Ambient light level controller. This illustrates an inverse application for the ABC sensor. In this case, the sensor is used to keep the ambient light environment at a constant level. As an example, lighting in a room can be increased in brightness to compensate for an outside decrease in sunlight. This is accomplished by driving a triac driver with the output of the ABC sensor. To obtain proper gating of the triac, the sensor must be synchronized with AC line voltage. The light level is established by setting a constant current through transistor Q_2. The ABC sensor will then set the duty cycle of the output at a level that will produce enough light to generate an I_p current at pin 1, which is equal to the current through Q_2.

7. Special Function Photo ICs

Conclusion

All the circuits necessary to provide automatic brightness control of most displays are available on a single silicon chip. The automatic brightness control, which has totem pole output, will source or sink 50 mA or 20 mA respectively, can be externally adjusted for sensitivity and is capable of both analog and digital output and synchronous or asynchronous modes of operations.

7.3 TRIAC DRIVER PHOTOSENSORS

The zero current crossing triac driver photo IC is designed to interface electric controls (DC) to AC loads. It can control AC loads up to 100 mA or provide trigger current of up to 100 mA for controlling power triacs that can, in turn, control higher current AC loads. Figure 7.8 shows a schematic of the circuit for controlling a resistive load up to 100 mA. The principal would be the same for controlling a power triac, except that the photo IC would be connected between the power triac trigger lead (gate) and one side of the AC line.

Figure 7.9 shows the time relationship of the source or AC line voltage, the triac and load current, the triac voltage, and the control pulse from the IRED. The load is resistive. When the control current for the IRED turns "on," the source voltage (less the small voltage drop across the triac) is applied to the load. When the con-

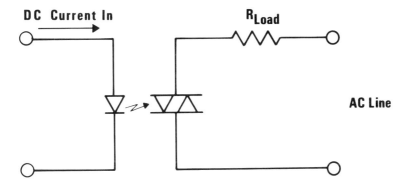

Figure 7.8 Electrical schematic of zero current crossing triac photo IC. The symbol for a triac is used for the photo IC because the device functions very much like a triac. At loads up to 100 mA, the device may be used in place of a triac, with IRED current substituting for gate current.

trol current for the IRED turns "off," the load current and voltage continue their sinusoidal path until the load current passes through zero. At that time, the triac turns off and the line voltage is dropped across the triac.

As the load becomes more inductive or capacitive, the turn on conditions remain the same, but the turn off condition that still occurs as the current through the triac goes to zero will cause a drop from the voltage across the load and triac to zero. As the load becomes inductive or capacitive, this voltage becomes peak line voltage. A "snubber" network may be added to minimize this effect on the turn off condition. Figure 7.10 shows the schematic of the photo IC triac or triac driver.

The zero current crossing triac driver photo IC still has the inherent disadvantage that the turn on point is not controlled. The turn on can occur when the line voltage is "high" across the load. The zero voltage crossing photo IC overcomes this disadvantage. The waveforms are identical to Figure 7.11, except the circuit does not turn on until the line voltage approaches zero. Figure 7.11 shows the schematic of the zero voltage crossing photo IC.

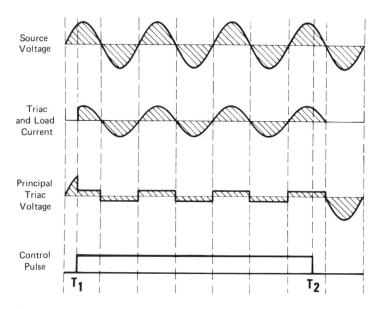

Figure 7.9 Principal voltage and current waveforms with resistive load. A resistive load indicates that voltage and current are in phase. Note that the device turns the triac on during a nonzero point in the voltage waveform. The off condition is reached as the current waveform reaches zero.

7. Special Function Photo ICs   123

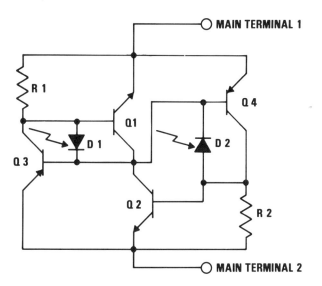

Figure 7.10 Electrical schematic of zero current crossing photo IC. Note the symmetry in the circuit. This, of course, is necessary to facilitate the AC functioning of the device.

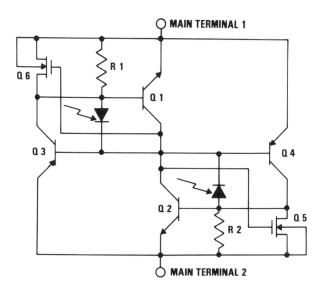

Figure 7.11 Electrical schematic of zero voltage crossing photo IC. The same symmetry for two alternate current directions is used as in the zero current crossing device; however, Q5 and Q6 are added to provice switching at zero voltage, eliminating the need for separate snubber circuitry.

III
THE COUPLED EMITTER (IRED) PHOTOSENSOR PAIR

8
The Transmissive Optical Switch

8.1 ELECTRICAL CONSIDERATIONS

The transmissive switch consists of an energy source (usually an IRED), an energy receiver (usually a photosensor), a means of interrupting the flow of energy and a means of fixing these in a mechanical configuration that allows repeatability of function. Part I covered the energy source, while Part II covered the energy receiver. This part covers the coupled pair in the transmissive mode.

Figure 8.1 shows a simple transmissive switch in the conducting (on) mode and the nonconducting (off) mode. When photosensor is conducting, the transmissive switch is in the "on" position. In order to understand the variables, a simple DC circuit will be used to illustrate the switch states of "on" and "off". In Figure 8.2, the simple DC circuit is shown, a load line on the V_{CE}/I_C characteristic curves, and an opaque member causing the switching action when it interrupts the energy beam.

The schematic of the IRED shows a current limiting resistor that essentially provides a constant current to the IRED causing a constant output of energy. The phototransistor effectively gives the switching action. When the energy path is blocked, the circuit's output current is equal to the leakage current or "dark" current, which is a characteristic of the phototransistor. The "off" voltage across the load resistor is equal to $I_C \times R_L$. This is normally very close to zero. The "on" voltage across the load resistor is $V_{CC} - V_{CE(SAT)}$. This is a function of "I_C" which is controlled by "h_{FE}." This is normally close to V_{CC}. The electrical "off" (low voltage across the load; high voltage across the transistor) and the electrical "on" (high voltage across the load; low voltage across the transistor) is clearly shown. For additional details on this, see Figure 5.9 and accompanying discussion in Chapter 5.

The transition time between "on" and "off," and "off" and "on" must be clearly understood with respect to the mechanical delays or changes in turn "on" and turn "off" in order to have a successful design. The most difficult portion to understand goes back to the effective blocking of the photosensitive area by the flag or blocking media. Assume an effective photosensitive area that is 0.0625 in. in diameter. Assume a

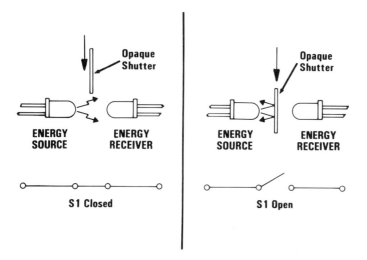

Figure 8.1 Transmissive switch. Optical motion sensing, in its most basic form, is illustrated by the simple transmissive optical switch. The position of the opaque shutter or "flag" determines whether the switch is open or closed.

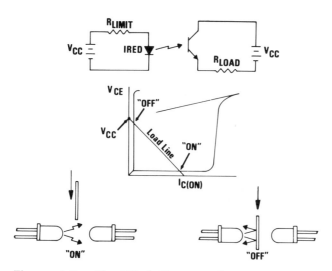

Figure 8.2 Simplified diagram of an optoswitch. Phototransistor collector current is predicted using the load line on the V_{CE}/I_C curve. The "on" condition drops almost all of the voltage across the load resistor and the collector current is at its maximum operating level for the circuit. The "off" state drops most of the voltage across the phototransistor at a low current level.

8. The Transmissive Optical Switch

blocking media passing between the energy source and the energy receiver, as shown in Figure 8.3.

Assume the following three sets of conditions:

1. "X" is larger than 0.0625 in.
2. "X" is equal to 0.0625 in.
3. "X" is smaller than 0.0625 in.

The blocking media is now moved between the energy source and receiver in a downward direction and the plot of voltage across the load resistor (circuit shown in Figure 8.2) versus distance is shown in Figures 8.4, 8.5, and 8.6.

As the opening of the blocking member starts to uncover the photosensitive area, the voltage across the load starts to increase. Once the opening fully exposes the photosensitive area, the voltage across the load reaches a maximum and remains there until the photosensitive area starts to be covered or blocked again. The voltage across the load resistance decreases until the photosensitive area is completely covered and remains there until the next opening starts to appear. The minimum voltage across the load is controlled by the leakage current (times the load resistance). If the sensor is in a dark environment the leakage current will be minimized. As the ambient light level is raised, this leakage current will increase.

The maximum value of voltage across the load occurs when the energy is adequate to drive the photosensor into saturation. If the energy is not sufficient for this to occur, the maximum value will drop, due to the fact that the transistor will not be in the saturated condition, but the relative times or slopes of the transitions will remain the same. The slope of the rising and falling load voltage will flatten out as the optical gain starts to decrease. As the optical gain starts to increase, the transition times will decrease and the voltage across the load will remain at the maximum for a longer time.

The width of "X" is now reduced to 0.0625 in. or equal to the diameter of the effective photosensor. The resultant voltage across the load versus the distance traveled is shown in Figure 8.5. Note the resultant triangular wave form.

As the optical gain of the system decreases, the voltage across the load will decrease. As the optical gain increases, then the transition times decrease and the maximum load voltage will decrease and the flat or saturated portion will increase. Since the photosensor is never turned completely off, the "off" condition will have a higher level. The "on" distance will remain the same.

The width of "X" is now reduced to less than 0.0625 in. The resultant voltage across the load versus the distance traveled is shown in Figure 8.6. Note that the rise and fall times occur in the "X" distance and the flat portion is 0.0625 in. $-X$.

Again as the optical gain of the system decreases, the maximum voltage across the load will decrease. The "on" distance will remain the

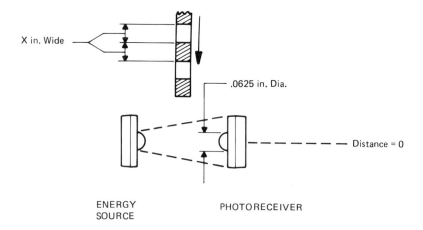

Figure 8.3 Energy blocking media. The width of the blocking media and aperture will affect the output states of the optical switch. Collector current and collector-emitter voltage will change accordingly as different amounts of energy are permitted to reach the photosensitive area.

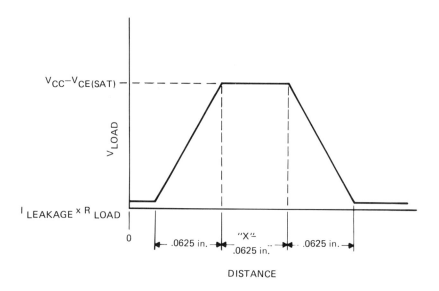

Figure 8.4 V_{Load} versus distance when "X" is larger than 0.0625 in. As the blocking media shown in Figure 8.3 passes through the beam, the sensing area becomes completely blocked or "dark." The load voltage is equal to $V_{CC} - V_{CE(SAT)}$ in this state.

8. The Transmissive Optical Switch

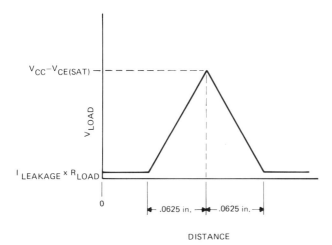

Figure 8.5 V_{LOAD} versus distance when "X" is equal to 0.0625 in. In this case, the sensing area is only completely covered at the instant that the blocking media is exactly centered. The load voltage still equals $V_{CC} - V_{CE(SAT)}$ but only manifests itself as the peak of the triangular wave.

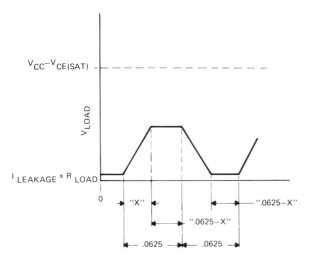

Figure 8.6 V_{LOAD} versus distance when "X" is less than 0.0625 in. The full "off" state is not reached if the slot width is less than the width of the sensing area. The width of the sensor and speed of the shutter determine the cycle time; however, the load voltage never rises to its possible maximum.

The Coupled Emitter (IRED) Photosensor Pair

same. At "X" distance from the trailing edge of $V_{LOAD(maximum)}$, the cycle repeats itself.

The discussion of mechanical rise and fall time is complete. Another variable can now be added into the system and understood. The switching time of the photosensor may also affect these curves. These curves are offset by the rise time of the photosensor (offsetting the turn-on condition) and the fall time of the photosensor (offsetting the turn-off condition). Figure 8.7 shows this phenomena. It is assumed that turn-on time for the photosensor (10 to 90%) is five μsec, and "X" is equal to 0.031 in. which corresponds to 10 μsec, for the assumed speed of travel.

In most designs, the mechanical turn-on and turn-off are significantly larger than the switching times of the photosensor. Figure 8.7 utilizes a combination of these switching times to illustrate the turn-on and turn-off times. The overall mechanical turn-on time is controlled by limiting the width of the opening in the opaque member passing the photosensor. The overall switching time of the photosensor is controlled by the selection of the photosensor. Review of

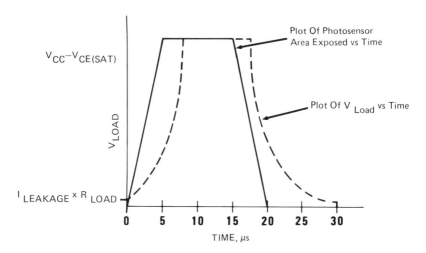

Figure 8.7 V_{LOAD} versus time when conditions are:

T_{ON} = 5 μsec T_{OFF} = 10 μsec 0.031 in. travel = 10 μsec

The rise and fall times inherent in the photosensor delay and alter the actual load voltage waveform when compared to the exposure versus time. This should be accounted for in designs requiring high speed switching (e.g., motor encoders).

8. The Transmissive Optical Switch

Section 2 on the receiver shows that a typical photodarlington (Figure 5.2) switches in milliseconds, a phototransistor (Figures 5.19 and 5.20) switches in microseconds, and a photo IC can approach the low nanosecond range.

The physical separation of the energy source and the photosensor, the effective radiating area of the energy source, the spread or scatter of the photons from the energy source, the effective photosensitive area, and the mechanical configuration of the interruptive medium contribute to the complexities of the optical switch.

Understanding how these variables interrelate is vital to a successful optical switch design. Figure 8.8 shows a photosensor with a convex lens mounted into a housing.

The energy received by the photosensor in Figure 8.8 is limited by the width of the housing slot at the most distant point from the photosensor. In other designs, the limit could also come from the photosensitive area on the chip or its optical projection caused by the convex lens.

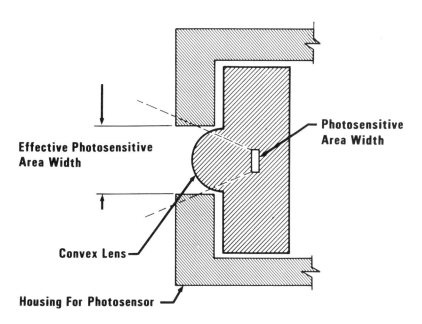

Figure 8.8 Mechanical outline for photosensor with convex lens mounted in housing. The primary function of the housing is to control the optical alignment and physical location of the sensor. Depending on material choice and design, the housing also serves to direct and control light and infrared energy incident on the photosensor.

The assumption is made that the opening in the opaque medium and the aperture in the interrupting medium in front of the photosensor are smaller than the effective radiating area of the transmitter. This is the normal case when time precision of turn-on or turn-off is required. Figure 8.9 shows the decrease in photosensor output as the opaque medium is moved from the transmitting side of the optical switch to the receiving side. The transmitter is radiating noncollimated energy. The receiver gets the maximum energy when the opaque member is against the transmitter and the amount of energy decreases as the opaque member moves toward the photosensor.

This can lead to varying pulse widths of the output signal if the opaque member does not follow a "true" path as it moves by the photosensor. Figure 8.10 illustrates this principle. The electronic sensing of the signal from the photosensor trips at a given signal level. As the opaque member moves back and forth in the gap, the trip point changes with respect to relative location of the sensor opening to the opaque member opening.

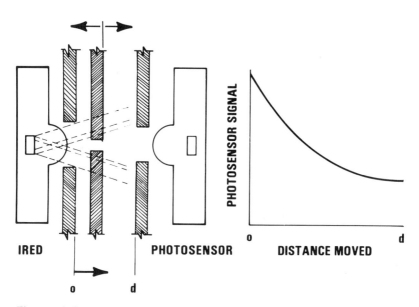

Figure 8.9   Decrease in photosensor output as opaque member moves from transmitter side to photosensor side. The diverging beam pattern means that flux density drops with distance. The greatest amount of energy passes through the opaque member when it is closest to the source.

8. The Transmissive Optical Switch   135

The net effect of this problem is identical to the effect discussed earlier on width of apertures (Figures 8.4, 8.5, and 8.6) versus optical gain. This is variation in optical gain caused by the location of the beam interrupt mechanism rather than the variation of optical gain from optical switch to optical switch caused by varying sensitivity of the photosensor or varying energy output from the transmitter.

Discussion up to this point has assumed that the energy interruption is accomplished with an opaque member. Many applications require the use of a semiopaque member to block the energy beam. The designer must first establish the transmission characteristics of the semiopaque member to the energy source and the photosensor. A simple technique is to use a PN photodiode (the collector-base diode of a phototransistor) and the IRED to be used in the final system. Table 8.1 shows typical readings taken on ten samples of a semiopaque member (paper). The IRED is biased to 20 mA and the photodiode current is read, with and without the paper samples interrupting the beam.

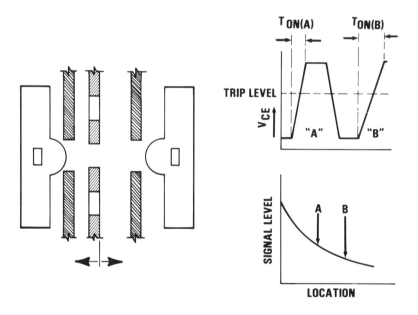

Figure 8.10 Change in trip points versus location of opaque member. The phenomenon illustrated in Figure 8.9 can alter the turn-on point if the opaque member (e.g., encoder wheel) physically drifts between emitter and sensor. The trigger voltage (for the circuit) is reached more slowly (B) as a result of the lower level of incident radiation on the phototransistor.

Table 8.1 $I_{C(ON)}$ Readings with and Without Paper Blocking Beam

Sample number	$I_{C(ON)}$ blocked	$I_{C(ON)}$ unblocked
1	2.4 µA	23.5 µA
2	2.3 µA	24.0 µA
3	2.35 µA	23.0 µA
4	2.45 µA	24.5 µA
5	2.5 µA	25.2 µA
6	2.45 µA	24.2 µA
7	2.35 µA	23.6 µA
8	2.5 µA	25.0 µA
9	2.3 µA	23.5 µA
10	2.4 µA	24.5 µA
Average	2.4 µA	24.0 µA

The paper is seen as 10% transparent or 90% opaque. If it is assumed that a 10% change in system operation is due to thermal changes, a 20% decrease due to IRED degradation and a 20% safety factor is required, then the unit to unit variation must be no more than five to one. Utilizing the system IRED drive the $I_{C(ON)}$ spread of the sensor must have a spread of five to one or less. These conditions may vary but the approach can be used to define the final specification.

Several other techniques have been used to solve the "transparency" problem. A potentiometer may be added to the IRED bias circuitry and adjusted on each assembly to a predetermined output level. This would overcome the unit to unit variation but would not overcome the thermal effects and degradation effects. A cheaper but less effective solution is to utilize an in-line-test that would vary the current limiting resistor in series with the IRED. This could be similar to the potentiometer approach but would utilize step variation rather than a linear variation. Another approach would be categorization of the IREDs in a mechanical mock-up of the system with a standard photosensor in grades of, say, 1 through 5, with 5 being the higher outputs. The photosensor could also be graded with a standard IRED

8. The Transmissive Optical Switch

of, say, 1 to 5 with 5 being the higher sensitivities. The unit could be mounted into the system paired 1 with 5, 2 with 4, 3 with 3, 4 with 2, and 5 with 1. This would tighten the distribution.

Another technique utilizes the nonlinearity of the photosensor's response versus energy. A phototransistor is less linear than a photodiode and a photodarlington is less linear than a phototransistor. If the photosensor is biased at low $I_{C(ON)}$ levels (where the nonlinearity is pronounced) then the signal differentials will be more pronounced.

Another technique utilized for minimizing the degradation effect of the IRED mounts a second photosensor near the IRED to monitor the output energy before it reaches the blocking material. This photosensor is used in a feedback circuit that changes the current through the IRED to give a constant level infrared output.

Another technique that is extremely effective when the interrupting media is very transparent is to cant or change the optical path to take advantage of the critical angle discussed in the section on IREDs. Figure 8.11 illustrates this phenomena. The energy path is bent by the interrupting media and deflected away from the sensor. This system is also used in the detection of bubbles where the leading edge of the bubble bends the energy away from the sensor.

Once the critical angle is reached, then all of the energy is reflected and none penetrates the interrupting media. Since the energy is deflected more as the angle increases, then less energy is available to the photosensor. The angle involved is the obtuse angle minus 90° (the energy path to the interrupting media). If mechanical restraints

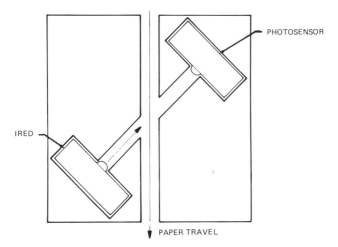

Figure 8.11 Utilizing the critical angle to deflect the energy beam. As the transmitter receiver plane is rotated, a portion of the energy is deflected away from the receiver improving the "on"/"off" ratio.

prevent the critical angle from being obtained, then the blocked/unblocked ratio can still be improved by as much canting of the paper with respect to the energy path as possible. (This canting may be in either a vertical or horizontal plane.)

Linear and rotary encoders have come in a wide variety of design styles over the years, the most common being rotary switches, potentiometers, capacitive, magnetic, and optical types. The optical encoder has become the most popular of these encoding methods due to its long life, simplicity of construction, versatility, high accuracy, and high resolution.

An encoder is an electromechanical device used to monitor the motion or position of an operating mechanism, and to translate that information into a useful output. An optical encoder is defined as an optoelectronic device that translates rotational or linear movement into some usable electronic waveform. Encoders generally consist of two parts. The first of these is a "moving" unit that is attached to and moves with the device being monitored. The moving unit contains information to be sensed by the "stationary unit." The stationary unit consists of an IRED and a photosensor (or a combination of IREDs and photosensors) mounted on opposite sides of a slot through which the moving unit passes, thereby modulating the light path(s).

The types of output information available are speed, velocity (speed with direction), and relative or absolute positioning. The output can be either analog or digital depending on the type of photosensor used.

Encoder Components

THE MOVING UNIT: The modulation of the light path(s) in the optical encoder is accomplished by the moving unit, which is a *scale* (linear encoder) or a *disc* (rotational encoder). The scale or disc is attached to the operating mechanism and contains alternating areas of transparency and opacity of the light path. The size, shape, and frequency of these areas is the basis of the output information supplied by the encoder.

A number of materials are currently being used in the fabrication of scale and disc components. A few examples are given in Figures 8.12, 8.13, 8.14, and 8.15, with the advantages and disadvantages of each.

THE STATIONARY UNIT: The stationary unit contains all the components necessary to generate the light source and sense its intensity as it is being modulated by the scale or disc. It sometimes contains the signal conditioning electronics required to amplify and/or digitize the output of the encoder. The light source consists of one or more incandescent lamps or light-emitting diodes and may include lensing to improve the collimation of the light source. Most recent optical encoders use IREDs because of their lower cost, longer life, better shock resistance, and lower power consumption.

8. The Transmissive Optical Switch

Advantages	Disadvantages
Low cost	Resolution of <50 lines
Durable	Relative mechanical and thermal instability

Figure 8.12   Molded plastic encoder disc.

1. Sensing Elements: Solar cells, photodiodes, phototransistors, and photosensitive integrated circuits are all used in optical encoders. A Photologic T.M. (trademark of TRW, Inc.) IC has been developed to enable the stationary unit to provide a digital output which can be directly interfaced with TTL, LSTTL, CMOS, and other standard logic families.
2. Apertures and reticles: One method of improving encoder resolution is the *sizing down* of the photosensitive area. This is done by placing a reticle with a certain aperture size in front of the photosensor. The reticle contains a pattern of transparent and opaque areas which are optically mated to the scale or disc

140   The Coupled Emitter (IRED) Photosensor Pair

Advantages	Disadvantages
Reasonable cost	Resolution of <150 lines per inch
Resistant to shock and vibration	
Good thermal stability	

Figure 8.13   Etched metal encoder disc.

being "read." The transparent areas are referred to as apertures, and one or more apertures may be placed in the reticle over the photosensor in high resolution designs. Some examples of reticles made of the same materials, and intended to be used with the scale and disc samples discussed earlier, are shown in Figure 8.16. The same advantages and disadvantages apply. In the case of molded plastic, apertures are molded right into the housing.

3. Signal conditioning electronics: Resistors, capacitors, integrated circuits, input/output connectors, and additional components are

8. The Transmissive Optical Switch 141

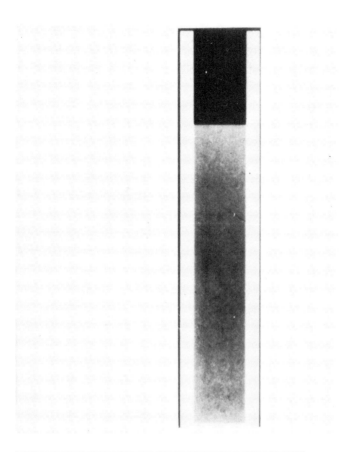

Advantages	Disadvantages
Reasonable cost	Mechanical, thermal, and humidity instability
Resolution of <100 lines/inch	Can be damaged in handling

Figure 8.14 Mylar film encoder.

142   The Coupled Emitter (IRED) Photosensor Pair

Advantages	Disadvantages
Resolution of >2500 lines/inch	High cost
Excellent optical quality	Can be damaged in handling
Excellent mechanical, thermal, and humidity stability	

Figure 8.15   Chrome on glass encoder disc.

   often contained on a printed circuit board in the stationary unit.
   These components are used to amplify the photosensor output and
   interface the encoder to the system in which it is used.
4. Housing: The components used in the construction of the stationary unit are usually held in position by mounting them into a
   metal or lastic housing. The housing is then mounted to the
   operating mechanism (motor, etc.) to optimize the interface between the moving and stationary units. In some cases, the moving and stationary units are packaged together and external link-
   ages are provided for coupling the packaged encoder to the operating mechanism.

8. The Transmissive Optical Switch 143

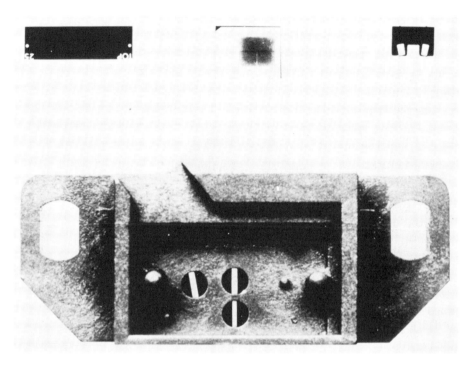

Figure 8.16 Examples of reticles. Reticles cover a full spectrum of materials and precision of manufacture. Their effective use follows a similar gradient from types requiring very little optical alignment to types needing very delicate alignment and handling.

Operating Principles

MODULATING THE LIGHT SOURCE: The movement of the scale or disc in the light path is the source of modulation of the light in an optical encoder. A simple example of modulation would be the interruption of the light beam in a burglar alarm. The momentary interruption or reduction of light is easily detected. As resolution requirements increase, apertures become smaller and detection becomes more difficult. An improvement over standard aperturing is the light shutter.

THE LIGHT SHUTTER: The reticles used in optical encoders may contain 20 or more alternating transparent/opaque areas in front of each sensor. If the moving unit and the reticle have identically matched patterns of 50% duty cycle (transparent and opaque areas are the same width) then the emitted light received by the sensor will be at a maximum when all the transparent areas of the reticle are exactly superimposed with those of the moving unit, as illustrated in Figure 8.17.

When the moving unit moves one area width, the emitted light received by the sensor will be at a minimum, but not zero, since in this type of light modulation, there is some slight light leakage around the opaque areas in the moving unit. This sequence repeats for each cycle of movement and is referred to as the "light-shutter" because of the similarity of operation to a camera shutter.

QUADRATURE: Determination of direction of movement of the moving unit is also possible by locating two photosensors in the encoder and mechanically shifting the aperture pattern in the reticle over one photosensor, one-quarter cycle from the aperture pattern in the reticle over the other photosensor as shown below. This causes a "phase shift" in the output of one photosensor relative to the other and indicates direction of motion. This phase relationship is called *Quadrature* and is illustrated in Figure 8.18.

The output from photosensor "A" rises 90° ahead of the output from photosensor "B," indicating that the moving unit is moving to the right. If the moving unit were moving to the left, the output from "B" would be 90° ahead of "A" (see Figure 8.19).

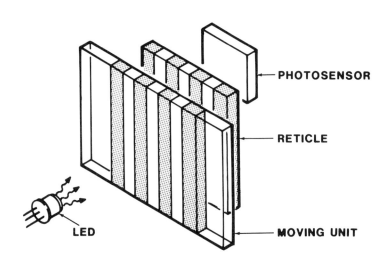

Figure 8.17 Example of light shutter. Most light shutters are designed with equal spacing of opaque and open areas. In actual use, there is a small but allowable amount of energy that still passes through the shutter when it is completely closed.

8. The Transmissive Optical Switch

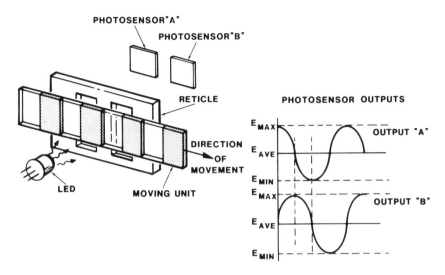

Figure 8.18 Example of quadrature. The use of quadrature allows the direction of rotation to be sensed. The determinant is whether the "B" output leads or lags the "A" output.

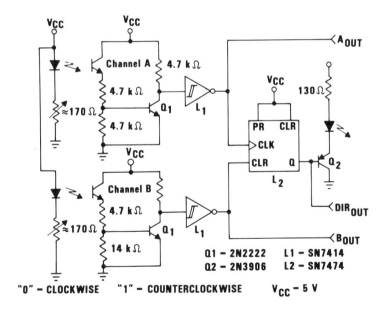

Figure 8.19 Schematic for determination of rotating direction, relative position, and speed. The circuit generates an output (Q), which signals clockwise rotation in the low ("0") state and counterclockwise rotation in the high ("1") state.

As shown in Figure 8.19, channel "B" provides the "D" input and channel "A" provides the "clock" input to the SN7474. (The SN7474 converts the relatively slow transitions from the mechanical motion to TTL compatible rise and fall times.) Since channel "A" clocks the latch at its positive transition, the state of channel "B" at "D" determines the state of the latch. If the "Q" output of the latch is high ("1" state) then the "D" input was high when channel "A" turned "on" prior to channel "B." This implies counterclockwise direction of rotation. For clockwise, "A" turns on prior to channel "B" and the "Q" output of the latch will be low ("0" state). The pulses at "A" out or "B" out may be used for speed and/or relative location. Speed may be determined by counting the output pulses for a given time period and dividing the total count by the number of pulses per revolution. Relative event location may be controlled by specifying the number of pulses between related events.

Sensing Circuit Techniques

The use of the light shutter permits the design of an optical encoder capable of very high resolution. However, electrical and mechanical errors must be considered and compensated for in the design to allow full use of this capability.

SINGLE-ENDED ENCODERS: The use of a single photosensor to generate each output in an optical encoder is inherently limited. IREDs will degrade with time and temperature resulting in changes in the output signal shape and level. However, if performance requirements are not severe, the single-ended approach offers the simplest design approach and lowest cost.

CONVOLUTED DUTY CYCLE ENCODERS: The use of 50% cycle components in a single-ended encoder does not necessarily guarantee the optimum in performance. A reduction in the duty cycle of the reticle (making the opaque area wider than the transparent area) and an increase in IRED drive current will improve the output performance of an encoder that is being digitized by a comparator. Operating a phototransistor at very high light conditions will tend to reduce its frequency response. The use of convoluted duty cycle usually requires the use of a photodiode type of photosensor. Again the photo IC series of photosensors is ideally suited for this type of application.

AUTOMATIC GAIN CONTROL: An unmodulated photosensor channel can be incorporated exclusively to monitor the intensity of the emitted light from the IRED. Feedback is then provided to a drive circuit powering the IRED. This compensates for degradation from all causes and will enhance the long-term performance of the encoder. The trade-off is in increased cost and circuit complexity.

8. The Transmissive Optical Switch

DIFFERENTIAL CIRCUITRY: By generating quadrature in "complementary format" (i.e., 0°/180°, 90°/270°), the complementary phases may be differentially amplified or compared to generate the required quadrature output (generally 0° and 90°). This approach allows noise reduction and drift compensation. An additional advantage is the ability to operate high gain phototransistors in the nonsaturated mode, thereby improving frequency response. The negatives are increased cost and circuit complexity.

ZERO REFERENCING: Many encoders provide speed, velocity, and relative position data, but a starting position must be known to derive true position. An extra photosensor is sometimes provided to look for a single point of transparency or opacity at a specific place on the scale or disc. The sensing of this point is used to zero the counting circuitry driven by the encoder durng power-up, or any time an error in count is detected.

Mechanical Interfacing

The best possible performance from an optical encoder is dependent on the proper selection of materials, circuit design, and the integrity with which the encoder is attached to the operating mechanism. The space between the scale or disc and the reticle must be as narrow as possible and consistently maintained throughout the travel of the moving unit. Variation will result in degraded performance.

MOUNTING THE MOVING UNIT: A properly designed housing provides for flatness across the surface of the reticle at some absolute height from the mounting surface of the stationary unit. This allows the positioning of the moving unit to be performed as a separate operation. Disc mounting requires two steps: (a) affixing the disc to a hub using adhesive and/or a clamp ring; and (b) mounting the hub/disc to the device being monitored using adhesives and/or set screws located 90° apart on the hub. Linear scales are mounted to a bracket on the operating mechanism at one or both ends. The entire scale must travel evenly and precisely through both ends extremes. A typical encoder mounting application is illustrated in Figure 8.20.

MOUNTING THE STATIONARY UNIT: The stationary unit should be designed to allow rotational or displacement adjustments. These adjustments compensate for mechanical tolerances in fabrication of the stationary and moving units that could prohibit the final fine tuning needed by the light shutter.

MAINTAINING THE GAP: The distance between the scale, or disc, and the reticle is referred to as the "gap." In photoemulsion type

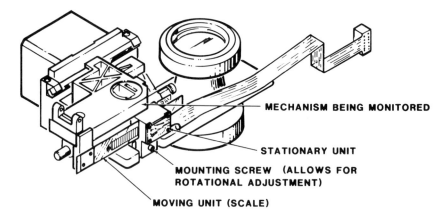

Figure 8.20 Typical encoder mounting application. This example is a linear motion encoder that might be used in industrial equipment, such as a lathe or milling machine.

light shutter components, the emulsion sides should be facing each other and a minimum space maintained to prevent abrasive damage. If the properties of the operating mechanism and the housing are known (thermal expansion, end play, eccentricity, etc.) the moving unit can be mounted using a spacer. Then the fixed unit is simply inserted between the shutter components to prevent wear damage.

ERROR RELATED TO THE GAP: A gap of zero width allows for complete modulation of the emitted light shutter. Any increase in gap width will result in reduced modulation where:

$$\% \text{ modulation} = \frac{\text{Signal output (ACVpp)}}{\text{Maximum achievable undistorted signal output}} \times 100$$

The reduced modulation is caused by noncollimated light from the IRED (i.e., leakage around the shutter components) and becomes substantial as the gap width approaches the aperture width in size.

Variations in the gap during the travel of the moving unit result in amplitude modulation. These variations affect the interface circuitry driven by the encoder during signal conditioning or digitizing and can cause clipping, positive pulse width modulation or variation in time between output pulses (in a pulse output encoder).

The quadrature relationship between the output channels will vary as the sum of the error on each individual channel.

PERFORMANCE LIMITS: The optical encoder provides direction information only as long as the quadrature related signals occur in proper sequence. Any phase, duty cycle, or modulation error that inter-

8. The Transmissive Optical Switch

rupts or reverses this sequence defines the ultimate limit of an incremental encoder.

Optical sensing is currently the most versatile method of motion sensing in rotary and linear applications. IRED and photosensitive integrated circuit technology, along with innovative sensing techniques, are keeping pace with today's sensing requirements so that the advantages of long life, high resolution, reliable operation in harsh environments, and low cost are available in almost any motion sensing application.

8.2 MECHANICAL CONSIDERATIONS

Housings for transmissive optical switches serve a dual role. They hold the discrete transmitter and receiver in a fixed location and also contain mounting holes or some other means of attaching the transmissive switch to the machine needing the switch function. The housing materials most widely used are a 10% glass-filled polycarbonate or polysulfone. These are used in an injection molding process to form the housing shape. Polysulfone has the unique property of transmitting wavelengths longer than 700 nm while blocking the shorter wavelengths. It is useful in housings that are used in a high dust ambient since there is no mechanical opening to clog up and thus block the optical path. The 10% glass-filled polycarbonate is used in most applications since it has higher solvent resistance and also blocks ambient energy in the long wavelengths. Other plastics are used for more specific applications.

Apertures are molded into the plastic. In the polycarbonate material, an opening or slit is left in front of the transmitter and receiver for the optical energy to pass. In the polysulfone, the opening is created by utilizing Snell's law for transmission of energy from one type of material to another. Figure 8.21 illustrates this principle.

The energy from the IRED will not penetrate the angular portion in front of the IRED. The critical angle of the air/polysulfone interface has been exceded by design. Apertures can be molded in either polycarbonate or polysulfone quite easily to a minimum of 0.010 in. widths or heights. Typically, the apertures are shaped in a rectangular form to allow the maximum transfer of energy while holding the aperture narrow to give maximum resolution. For resolution below 0.010 in., a cost versus accuracy trade-off comes into play. A separate aperture (usually etched from a thin metal) can give accurate definition but adds cost due to the extra piece part and the labor cost to install it in the optical switch housing. The molded aperture is less costly but starts to lose definition and accuracy due to accumulated tolerances. Figure 8.22 shows the detail of a molded aperture that is 0.010 in. × 0.050 in. and an etched aperture that is 0.007 in. × 0.040 in.

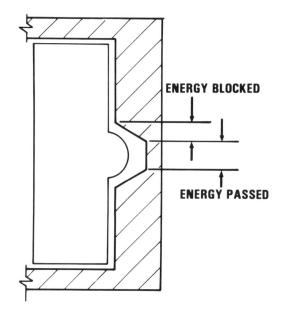

Figure 8.21 Polysulfone aperture design. Polysulfone passes wavelengths greater than 700 nm and may be used to construct a housing with excellent resistance to dirt and dust.

Figure 8.22 Molded versus etched aperture. Molding is generally the manufacturing method of choice on sizes down to 0.010" due to its low cost. Smaller apertures are often etched.

8. The Transmissive Optical Switch   151

The normal units utilized in these optical switches emit and receive energy perpendicular to the egress leads. A photograph of an emitter/sensor pair is shown in Figure 8.23. The exact shape differs among manufacturers but the function is essentially the same.

This technique facilitates mounting in either printed circuit boards or sockets and simplifies the design for space considerations of the molded holder. The lens design on this type of package is inefficient due to the close proximity of the semiconductor chip and the lens. As a result, most optical switches utilizing these discretes

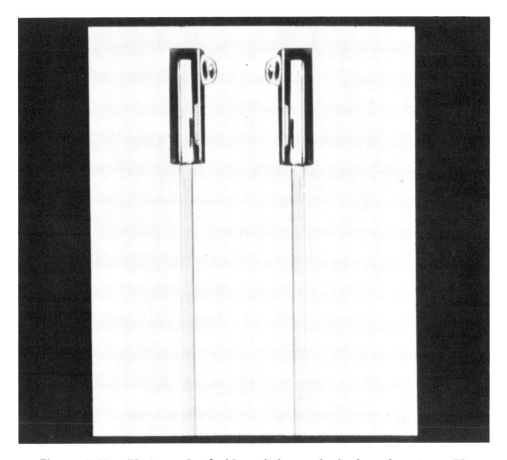

Figure 8.23 Photograph of side emitting and viewing discretes. The energy is emitted and received about an optical axis perpendicular to the leads. As such, these devices are excellent for use in slotted switch design. Easy PC board mounting, wire attachment, or socket compatability is possible.

152   The Coupled Emitter (IRED) Photosensor Pair

Figure 8.24 Assorted types of standard optical switches are available from various manufacturers.

Table 8.2 Reasons for Deviating from Standard Design

---
1. Change in housing material
2. Change in housing shape (mounting, aperturing, number of channels)
3. Addition of egress wires
4. Addition of connectors
5. Added electronics (IRED current limiting resistor, amplification electronics, wiring interconnects)
6. Tightened electrical parameters
7. Unique application requiring special technique
---

8. The Transmissive Optical Switch

are restricted to 0.125 in. or less air gap between the IRED and photosensor. Figure 8.24 is a photograph of a variety of standard available optical switches.

In many cases, the optical switch function or mounting is not compatible with the standard parts. Some manufacturers cater to this custom design. The custom design usually involves one or more changes from the standard package. A list of normal type changes is shown in Table 8.2.

The manufacturers who cater to this custom design will usually absorb the engineering design costs and pass on the tooling costs (molds, special or unique engineering fixturing) to the customer. Due to the fact that the cost of a simple mold is in excess of $10,000, the volume of parts produced generally must excede 10 to 25 K/yr to justify the tooling. The application of these custom parts are thus limited to large volume applications.

9
The Reflective Optical Switch

9.1 ELECTRICAL CONSIDERATIONS

Reflective switches may be divided into two separate categories. Surface *detectors* will be discussed in some detail, while *mark sense*, or line *detectors* will be discussed in much less detail. The basic reflective switch has both the transmitter and the receiver mounted together on the same side of the surface or line to be detected. Most units operate on a combination of diffused reflectance and direct reflectance. Figure 9.1 shows these two principles.

A highly polished surface (such as a mirror) will yield a high direct reflectance of energy, while a surface with diffused reflectance (such as white bond paper) will have a high diffused reflectance component of energy. In order to take advantage of these components, the design of the reflective switches is different for each case. The unit optimized for direct reflectance usually has the emitting and receiving elements mounted along the legs of an isosceles triangle with the reflective surface placed at their extended intersection. The unit optimized for diffused reflectance would mount the sensing and emitting elements parallel to each other with the reflectance surface perpendicular to their extension. Part I discussed the dispersion of energy from the IRED as the distance increased from the unit. Figure 9.2 shows this graphically for a convex lens device. Note that the reflective surface maintains the dispersion of energy such that the actual beam appears to follow the inverse square law relationship. Thus the actual energy the sensor sees in the reflective mode is a factor of four rather than two lower than the energy received at the reflective surface. Since the decrease of energy per unit area versus distance actually deviates from the inverse square law relationship (the energy source at close distances), the basic inverse square law assumption is technically false but is good as a first order approximation.

This is brought about by the reflecting surface further dispersing the energy. As the surface becomes less polished, then the direct reflected component will decrease and the diffused component will increase. Figure 9.3 shows a direct reflected unit designed to intersect at 0.200 in. from the front. Due to the diverging beam pattern

9. The Reflective Optical Switch

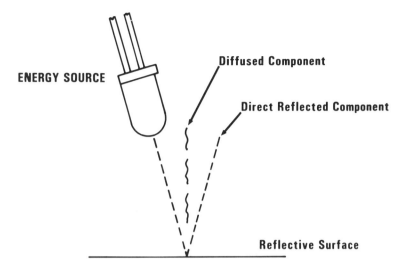

Figure 9.1 Diffused and direct reflectance. The angle of the direct reflected component is equal to the angle of incidence, while the diffused component could be directed virtually anywhere above the surface.

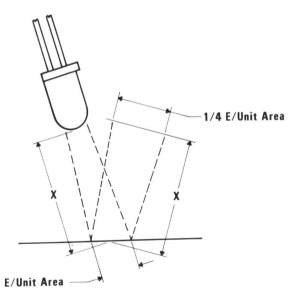

Figure 9.2 Graphical representation of reflected energy versus distance. A good approximation of energy level of reflected infrared is to use the inverse square law. Note, however, that diffusion and beam irregularities are not accounted for by this mode.

from the IRED, the actual peak response occurs at 0.150 in. This is due to the fact that the diverging pattern of the energy from the IRED and the shorter optical path the inside portion of the energy has to travel cause higher outputs before the actual mechanical peak is reached.

A unit designed for diffused reflectance usually is fabricated with devices having plano lenses. This reduces the efficiency of energy transfer but allows a flat surface (nondust-collecting) on the face of the unit. The peak response normally occurs close to the face of the unit (approximately 0.050 in). This is shown in Figure 9.4.

In both cases, the effective photosensitive area detects the energy reflected back from the reflecting surface. The amount of energy the sensor receives is a function of the reflectance of the reflective surface with respect to the projected sensor receiving area, the distance from the sensor to the reflective area, and the perpendicularity of the reflective surface to the plane through the transmitter and receiver. Table 9.1 shows relative readings from a direct reflectance unit and a diffused reflectance unit with respect to different types of surfaces. Analysis of the data shows several interesting points.

1. Surfaces 1 and 2 show the best response to the direct reflectance unit.

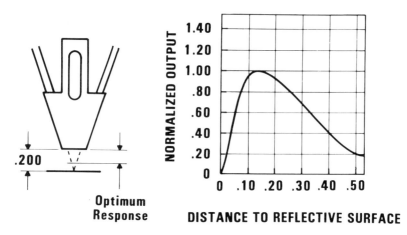

Figure 9.3 Relative output versus distance for direct reflectance. This curve is critical to the engineer. Optimum response is obtained by designing the reflective surface to pass about 0.150 in. from the switch housing.

9. The Reflective Optical Switch

2. Surface 4 shows the best response to the diffused reflectance unit. The diffused reflectance unit also is a good detector for the direct reflected energy.
3. The black velvet paint in surface 3 gives excellent contrast for either unit.
4. The signal level drops off on the direct reflectance unit but is quite good on the diffused reflectance unit when using neutral white bond (5) or mylar magnetic tape (7).
5. The graphite pencil mark in surface 6 gives relatively little change to the direct reflectance unit, since it both improves the reflectivity (smeared shiny graphite) and disrupts the pore fibers of the paper. The disruption of the pore fibers is more pronounced on the diffused reflectance unit.
6. Plastic surfaces do not reflect as well as polished surfaces (8, 10, 12, 14) on the direct reflectance unit but reflect almost as well on the diffused reflectance unit.
7. Plastic matte surface reflects almost as well as plastic smooth surfaces (9, 11, 13).
8. Color makes very little difference (8, 10, 12, 14, and 9, 11, 13).
9. The 3M tape #476 is an excellent nonreflecting surface (15).
10. The best nonreflecting surface is no surface (16).

The reflective sensor will normally work better than the transmissive sensor in three types of applications. The first of these is obviously when access can be had to only one side of the object to be detected. The second of these is in liquid level detection.

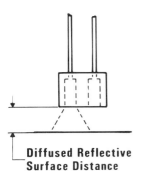
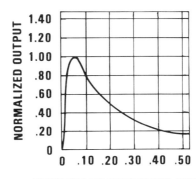

Figure 9.4 Relative output versus distance for diffused reflectance. As in Figure 9.3, the output versus distance curve is the key to a successful design. In this case, an unfocused reflective switch is used to detect a diffuse surface (reflective, but not highly polished).

Table 9.1 Reflectance versus Surface

OPB 253A $D^{(a)*} = 0.20$ in.	OPB 706 $D^{(a)*} = 0.05$ in.		Surface
689 μA	1000 μA	1.	Aluminum foil tape (shiny, efficient reflective surface).
680 μA	960 μA	2.	Alzak (similar to 1).
1.59 μA	20.99 μA	3.	Alzak painted with flat black velvet (3M #101-C10 black). Painted surface destroys shiny reflective surface and gives velvety matte finish.
115 μA	1950 μA	4.	Kodak 90% diffuse reflectance neutral white paper.
84.5 μA	860 μA	5.	White bond paper.
51.18 μA	390 μA	6.	No. 3 graphite on white bond paper with entire viewing area of sensor shaded by graphic mark.
41.9 μA	94.66 μA	7.	Mylar magnetic tape.
123.0 μA	964 μA	8.	Clear, smooth plastic tape finish.
90.0 μA	913 μA	9.	Same as (8), except matte color.
118.0 μA	985 μA	10.	Same as (8), except blue color.
100.0 μA	961 μA	11.	Same as (10), except matte finish.
116.0 μA	996 μA	12.	Same as (8), except red color.
106.0 μA	972 μA	13.	Same as (12), except matte finish.
107 μA	940 μA	14.	Same as (8), except gray color.
1.39 μA	24.20 μA	15.	3M tape #476 (a dull black surface).
0.24 μA	0.751 μA	16.	No reflective surface.

*(a) = distance from face of reflective unit to surface.

9. The Reflective Optical Switch

Figure 9.5 shows an outline of the principle of operation for a liquid level detector. This is more precise than a transmissive sensor and also allows the electronics to be remote with respect to the liquid.

The principle of operation is quite simple. The IRED energy normally is deflected by the angle at the bottom of the rod and returns to be detected by the photosensor. When a liquid such as gasoline covers the reflecting surface, the energy is not reflected but continues its direct path into the liquid (Snell's law when the N_2 interface changes from air to liquid). The third application involves precise height checking where the width across the material is very large compared to the distance from the detector to the material. Figure 9.6 illustrates this principle.

The application might require a tray or holder adjustment to paper checks or other light, small objects being automatically processed. The distance they dropped could be held to a minimum to prevent misorientation. As the number of checks in the holder is increased, the distance between the checks and reflective sensor would decrease. The sensing electronics would get an increasing signal. Once the checks get close enough to the sensor such that the signal started to decrease (see Figure 9.3), the electronics would trigger the check holder to drop or unload, increasing the distance, and starting the cycle over.

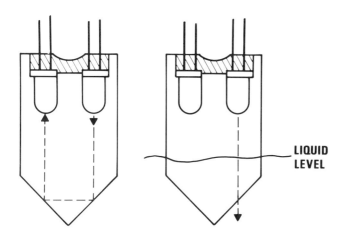

Figure 9.5 Liquid level sensor. The basic design for a liquid level sensor may be expanded to include multiple reflective steps to detect different fluid levels. System electronics would correspondingly increase in complexity.

The Coupled Emitter (IRED) Photosensor Pair

Mark sense or line detectors are optically much more complicated. In the previous discussion, it was pointed out that the area ratio between the effective viewing area of the photosensor and the area to be detected was a controlling variable. Assume an effective viewing area of the photosensor as a square that is 0.250 in. on a side. Assume a mark that is 0.005 in. side by 0.125 in. high with 100% reflectance. The reflective mark would give a signal change of 1%. It is quickly apparent that mark detection requires the system to become more complicated. Normally, both the collimation of the energy source and limitation of the effective photosensitive area is added to the mark sense or line detector. Three techniques are used to better collimate the energy source. Changing the source from a noncollimated IRED to a laser will dramatically improve the desired signal level. Adding a converging lens to better collimate the IRED energy will also improve the signal level. This may be done on both the transmitter and receiver side. Utilization of a light pipe as shown in Figure 9.7 will decrease the energy loss and can also effectively reduce the effective viewing area of the photosensor.

The basic approach that the design engineer must take when approaching a design requiring an optical switch is quite clear. First, attempt to select a transmissive switch. When that approach is unsuccessful, then consider the reflective switch. The additional variables and poorer resolution make the reflective switch the last choice.

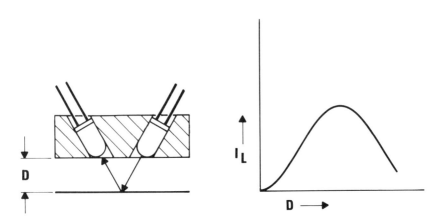

Figure 9.6 Optical technique used to reject reflections from close objects. At a certain distance, the output signal becomes large enough to trigger secondary activity, such as unloading a tray or moving a part.

9. The Reflective Optical Switch

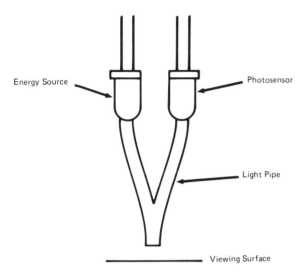

Figure 9.7 Utilization of a light pipe to improve signal level. The use of optical light pipes (e.g., fiber optics) offers the most flexibility in directing the energy to the surface and back.

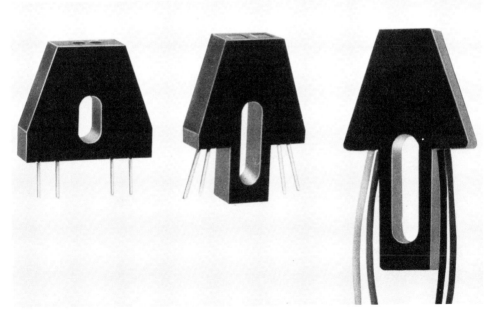

Figure 9.8 Photograph of focused reflective optical switches. As was the case with transmissive switches, a variety of reflective optical switches are commercially available as standard products. More complex applications often require customized designs.

9.2 MECHANICAL CONSIDERATIONS

The housings for reflective optical switches serve the same dual role as transmissive optical switches. They not only serve a mechanical function of locating the transmitter and receiver but also have a fastening system to the system or subsystem. The focused reflective assemblies normally take the shape of an arrowhead. A mounting slot is provided that provides adjustment of focus for optimum signal. Figure 9.8 shows a photograph of three different types of reflective units.

These are usually fabricated with coaxial discrete emitters and sensors (egress leads in same plane as the lens) with convex lenses to improve signal levels. Most of these units are not provided with dust covers. The dust cover with its finite thickness provides a reflecting media that introduces halation generated cross talk or noise into the system (see Figure 9.9).

The unfocused reflective sensor is fabricated from either discrete plano lens devices or an individual header or holder with the discrete

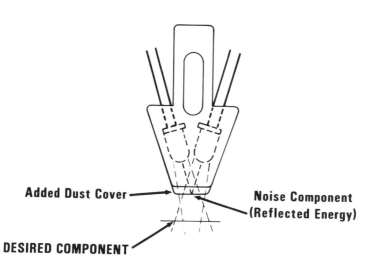

Figure 9.9 Focused reflective sensor with external dust cover. Halation from the internal dust cover reflections must be controlled or adjusted for by the signal processing circuitry before the design can work reliably. Similar problems have been known to occur when the reflective switch is mounted behind a glass window.

9. The Reflective Optical Switch

IRED and phototransistor chips mounted on the same header. Figure 9.10 shows a photographs of three such sensors.

The two sensors on the left and right are made with discrete transmitters and receivers while the one in the middle is fabricated from chips mounted on a TO-18 outline header with a molded plastic insert added, which forms a container for the epoxy protecting the chips as well as an energy blocking media between the chips.

Since these are made with plastic material, a dye can be added during fabrication. This dye effectively blocks all wavelengths shorter than 700 nm and passes the longer wavelengths of IR energy. The flat surface prevents the buildup of dust. The major drawback of this type of assembly is the difficulty in mounting. They may be mounted in a molded or machined cavity or also mounted by their leads in either a socket or on printed circuit board.

Figure 9.10 Unfocused reflective optical switches. The square shapes and parallel leads of these devices are due to the flat lensed discrete emitters and sensors (or chips) mounted on parallel optical axes.

IV
THE OPTICAL ISOLATOR OR COUPLER

10

Electrical Considerations

10.1 BACKGROUND

The optical isolator or optical coupler is similar to a transformer in that the input is electrically isolated from the output. The optical isolator or coupler is normally driven from a DC source and the output drives an electronic or DC circuit. Special cases of these couplers are available that operate from an AC source or drive an SCR or triac to control an AC load. In principle, they are very similar to the transmissive switch discussed in Part III. The air gap is replaced by a transparent material providing coupling of the IRED energy to the photosensor. Most applications utilize the units in the digital mode taking advantage of the isolation of signal from input to output. A portion of Chapter 10 will discuss utilization of the coupler in a linear mode and how to take advantage of this type of operation.

Figure 10.1 shows the construction of a coupler utilizing a discrete packaged IRED and photosensor mounted in a housing and separated by a light pipe. The physical separation of the source and receiver provides the electrical isolation of input to output. This obviously will vary as a function of the distance and the material separating the units electrically.

Table 10.1 lists currently available couplers by package, isolation voltage, and output function. Since the coupler resembles a transformer in function (providing electrical isolation), various agencies control some of the electrical and mechanical specifications. If the coupler were to short input to output, there would be a possibility of damage to equipment or personnel operating the equipment. This possibility of damage obviously varies from country to country due to the fact that one of the major causes of overload is voltage spikes or surges. The quality of the line voltage varies throughout the world. The amount of overload protection must then vary in accordance with electrical service provided and by the degree of conservation that exists within the controlling agency. Table 10.2 lists the major agencies and their type of control.

There are a number of other agencies but they exercise no control over the construction or characteristics of the coupler.

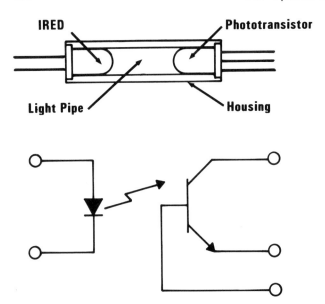

Figure 10.1 Optical isolator utilizing discrete transmitter and receiver. Perhaps the simplest design for an optically coupled isolator is to place discrete axial components at opposite ends of a hollow tube. The early designs, and some of the most popular designs, today, are based on this method of manufacture.

Table 10.1 Listing of Currently Available Types of Couplers

Package	Isolation voltage	Output
Coaxial discretes	10 to 50 kv	Photodiode transistor, Darlington, and IC
TO-18, TO-5	1 kv	Photodiode, transistor, Darlington
6-8 pin DIP	1.5 to 7 kv	Photodiode, transistor, Darlington, IC, SCR, triac driver
PC board mounted	6 kv	Phototransistor
Rectangular case		Photodarlington

10. Electrical Considerations 169

Table 10.2 Factors Controlled by Agency

Agency	Factors controlled
UL (USA)	Isolation voltage
VDE (Germany)	Egress terminal spacing and isolation voltage in three levels:
TUV (Germany), independent agency	1. Air conditioned office environment 2. Commercial environment 3. Industrial environment
CSA (Canada)	End equipment only
BSI (Britain)	Similar to CSA but accepts VDE standards

10.2 FUNCTION

In order to understand the function of the coupled pair, the input must be examined first, then the output, so that the two can be put together in one element. Figure 10.2 shows the energy output waveform from an IRED with a trapezoidal input waveform. The assumption is made that the turn-on and turn-off times of the IRED are negligible.

If the input waveform was sinusoidal, then the output waveform would be similar, with the IRED not turning on until approximately 0.9 V and turning off when the input voltage dropped below 0.9 V. Figure 10.3 shows the same type of plot when two back-to-back IREDs are used at the input.

There are several inconsistencies that should be considered when examining the effect of Figures 10.2 and 10.3. The energy output is not linear with increasing voltage or increasing current. The energy output is inconsistent from one IRED to another. The energy output must be offset by the turn-on and turn-off times of the IRED.

Now apply the output from Figure 10.3 to a photodiode and to a phototransistor. Figure 10.4 shows this relative relationship. R_L has been adjusted in magnitude so that output voltage is identical.

Without the adjustment in R_L, the voltage across the load for the photodiode would be much lower. It would approximate the voltage across the transistor R_L divided by h_{FE}. The curves must also be offset by the turn-on and turn-off times of the photosensor. The important point is the nonlinearity of gain versus current on the phototransistor. Figure 10.5 shows the the electrical schematic of a coupler with phototransistor output.

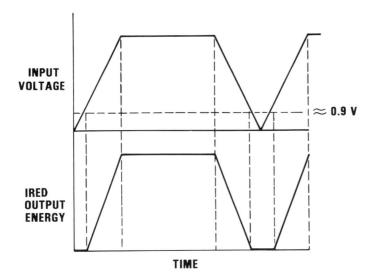

Figure 10.2 IRED output waveform versus trapezoidal input waveform. The infrared output of the diode tracks the voltage waveform above 0.9 volts. Below 0.9 volts, no infrared is emitted.

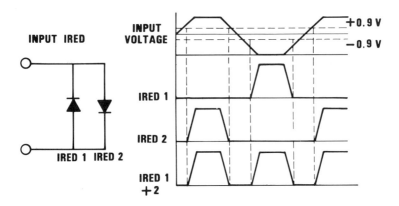

Figure 10.3 IREDs output waveform versus trapezoidal input waveform. Two IREDs in parallel, at opposite polarities, are used in AC input optocouplers. A typical application would be a telephone ring detector for an answering machine or computer.

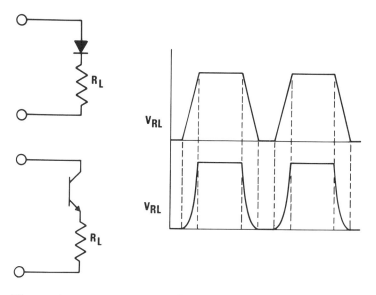

Figure 10.4 Output waveform from photodiode and phototransistor with input waveform from Figure 10.3. The linear response of the photodiode (top waveform) results in a waveform identical to the IRED voltage waveform. The nonlinearity of the phototransistor (lower waveform) introduces some distortion.

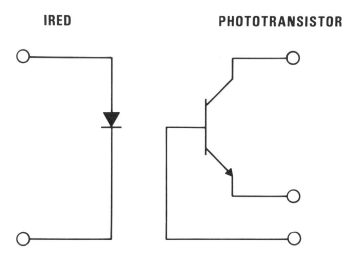

Figure 10.5 Electrical schematic of coupler with phototransistor output. The vast majority of couplers produced follow this basic schematic. Most are packaged in simple six pin plastic dual-in-line packages (PDIPs).

Table 10.3 Phototransistor Output and h_{FE} versus Input

IRED	$I_{TRANSISTOR}$	h_{FE}(XSTR)	Equivalent I_b
2	0.83 mA	335	2.5 µA
4	1.75 mA	348	5.0 µA
6	2.73 mA	364	7.5 µA
8	3.8 mA	380	10.0 µA
10	5.0 mA	400	12.5 µA
12	6.0 mA	400	15.0 µA

Assume that the output of the IRED is linear from 2 mA to 20 mA. Table 10.3 shows the output of the phototransistor versus the step input from the IRED and the subsequent calculation of h_{FE}. The equivalent I_b shows a linear input to the phototransistor.

Typically, a phototransistor is designed to be reasonably linear in the range of 500 µA to 20 mA. The h_{FE} becomes more nonlinear above 20 mA due to current crowding. The nonlinearity in h_{FE} below 500 mA is simply due to nonlinearity in h_{FE} efficiency. Analyzing this effect on the operating life of the coupler shows an exaggerated fall-off. Table 10.4 shows this effect on transistor output.

The measured degradation is worse than the assumed degradation. This is an unrealistic example due to the exaggerated assumed degradation, but the point is valid. The measured degradation will be exaggerated by the decrease in h_{FE} as the base current decreases.

The six pin plastic dual-in-line package (PDIP) is widely used for its low cost, dual sourcing, and performance characteristics. The most popular type is the phototransistor output with an electrical schematic, as shown in Figure 10.5. Many design engineers add a resistor (base to emitter) in order to improve the fall time. Figure 10.6 shows the switching time of a typical unit versus the value of base resistance.

Table 10.4 Exaggerated Operating Life for Coupler Degradation

IRED	Assumed degradation	New I_b	New I_c	Measured degradation
10 mA	0%	12.5 µA	5 mA	0%
10 mA	20%	10.0 µA	3.5 mA	24%
10 mA	40%	7.5 µA	2.73 mA	45%

10. Electrical Considerations 173

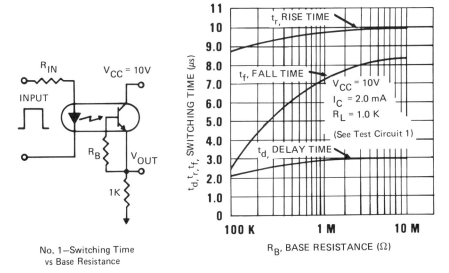

No. 1—Switching Time vs Base Resistance

Figure 10.6 Delay, rise, and fall time versus base resistance. The base resistor facilitates the tradeoff possible between collector current and switching time. Any current that passes through the base resistor must be subtracted from the photocurrent.

Assume a bases resistance of 200 KΩ. This would improve the fall time from approximately 8 µs to 4.4 µs. The normal base to emitter bias level would be approximately 1 V. This means that 5 µA of base current would be required for the 200 Ω resistor. This current would have to be subtracted from the photocurrent of the phototransistor. If we now repeat the data of Table 10.3 as Table 10.4, we see an IRED current of 10 mA will only generate 2.73 mA of phototransistor collector current.

The effective transistor h_{FE} has dropped from 400 to 218. Now assume a recalculation of Table 10.4. In Table 10.5, which is a recalculation of Table 10.4, we again assume an IRED degradation of 20% and 40%.

A 20% IRED degradation now shows 36% coupler output degradation rather than 24%. A 40% IRED degradation now shows a 67% coupler degradation rather than 45%. Even though the values are exaggerated the trend is definitely there. The 5 µA that is required to satisfy the current drain requirements of the 200 KΩ base resistor significantly effects the apparent degradation of the coupler. This factor must be included in the design engineer's calculation for long-term system performance. Note that the degradation of 20% of IRED output causes a 24% degradation in sensor output due to h_{FE} change and a 36% change when a base emitter speed-up resistance is added.

The performance of the coupler versus temperature generally will track the performance of a reflective or transmissive switch versus

Table 10.5 Exaggerated Operating Life for Coupler Degredation with R_B of 200 KΩ.

I_{LED}	Assumed degradation	New I_b	New I_c	Measured degradation
10 mA	0%	7.5 µA	2.73 mA	0%
10 mA	20%	5 µA	1.75 mA	36%
10 mA	40%	2.5 µA	0.83 mA	67%

temperature. Normally a coupler has the transmitter and receiver located in closer proximity to each other and, as a result, the photosensor and the IRED operate at closer to the same junction temperature. Figure 10.7 shows the relative change in output for an IRED and a phototransistor. Figure 10.8 shows their combined characteristics.

The IRED curve is reasonably constant independent of output. As the phototransistor increases in h_{FE} the upward slope will become pronounced. As the phototransistor decreases in h_{FE}, the slope will turn downward. Note that the phototransistor dominates at low temperatures while the IRED dominates at high temperatures.

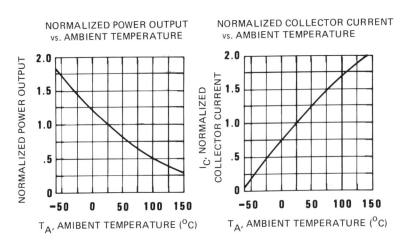

Figure 10.7 Normalized output versus ambient temperature for an IRED and phototransistor. The IRED output and collector current changes with respect to temperature are opposite functions. Increasing temperature results in less infrared energy, but simultaneously, the phototransistor becomes more sensitive.

10. Electrical Considerations

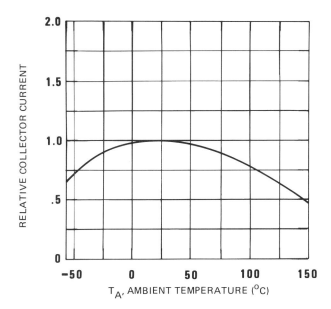

Figure 10.8 Relative collector current versus ambient temperature for a typical optocoupler. At temperature extremes, one effect always dominates the other. Below 25°C the low transistor gain takes over. Above 25°C the low IRED output dominates. Optimal operation takes place in the 0 to 50°C range.

10.3 DIFFERENT TYPES

Optocouplers may be divided into five basic types: standard six pin PDIP, AC couplers, speciality, hermetic or military, and high technology couplers. The standard type includes the six pin PDIP. An outline drawing and photograph is shown in Figure 10.9.

These units are available from a variety of manufacturers. JEDEC (Joint Electronic Device Engineering Council) registration has been made on a series of couplers with transistor and darlington outputs. Table 10.6 shows a listing of these parts.

In addition to these JEDEC registered parts, each manufacturer has a variety of standard couplers with varying specifications listed under their own unique part numbers. The fabrication techniques for these couplers are covered in Section 11.1.

The AC couplers consist of three different types. The major volume parts consist of an IRED input coupled to an SCR or triac driver output. These devices are very useful in controlling AC loads by DC electronics. The most popular types are the IRED input coupled to either a zero current crossing triac driver switch or a

176   The Optical Isolator or Coupler

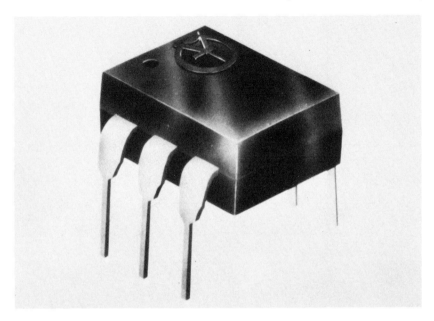

(a)

Figure 10.9   Photograph and outline drawings of six pin PDIP. This standard design has become the industry's most popular optocoupler package. Many output types are possible, including complex IC outputs.

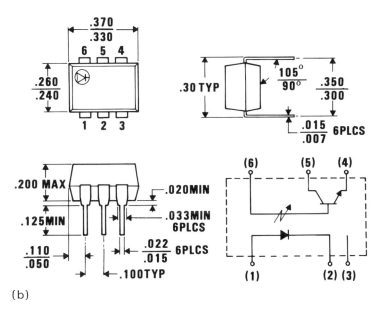

(b)

Figure 10.9   (continued)

Table 10.6   JEDEC Registered Six Pin PDIPS

Number	Isolation	Current transfer ratio	Output
4N 25	2500VDC	20%	XSTR
4N 26	1500VDC	20%	XSTR
4N 27	1500VDC	10%	XSTR
4N 28	500VDC	10%	XSTR
4N 29	2500VDC	100%	DARLINGTON
4N 30	1500VDC	100%	DARLINGTON
4N 31	1500VD	50%	DARLINGTON
4N 32	2500VDC	500%	DARLINGTON
4N 33	1500VDC	500%	DARLINGTON
4N 35	2500VRMS	100%	XSTR
4N 36	1750VRMS	100%	XSTR
4N 37	1050VRMS	100%	XSTR
4N 38	1500VRMS	20%	XSTR
4N 38A	1775VRMS	20%	XSTR

zero voltage crossing triac driver switch. The third type (the inverse) operates from an AC input to a DC load. Figure 10.10 shows the schematic for currently available couplers of this type.

Specialty couplers are also available from a variety of manufacturers. They generally differ significantly in shape. They feature high isolation voltage brought about by the physical shape of the parts. Figure 10.11 shows a coaxial part with 10 kVDC isolation. Figure 10.12 shows another coaxial part with 15 kVDC isolation, and Figure 10.13 shows one with 50 kVDC isolation.

Figure 10.14 shows a slightly different packaging approach using a side viewing discrete sensor and side emitting idscrete IRED mounted in a molded package.

Hermetic or high reliability couplers are designed such that they can be processed and tested to a military type specification such as Mil-S-19500/486A or Mil-S-883 Class B. These are available in the TO-18 outline, the TO-5 outline, and the hermetic DIP outline packages. Some manufacturers have in-house high reliability processing to augment the military approved parts. This allows the customer to generate a military type specification on a coupler that has not had a specification released by the military.

As the name implies, high technology couplers are more complicated than the types of couplers previously discussed. They are IRED or DC input combined with a photosensitive integrated circuit output. The 6N135 and 6N136 are graded on current transfer ratio and are the simplest form of a photo IC. The electrical schematic is shown in Figure 10.15.

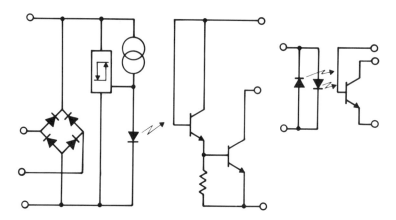

Figure 10.10   Optocouplers operating from an AC input to a DC output. In many applications, an AC signal requires monitoring. The bridge type and back-to-back designs are both effective methods of AC/DC conversion; however, the back-to-back method is more popular.

10. Electrical Considerations

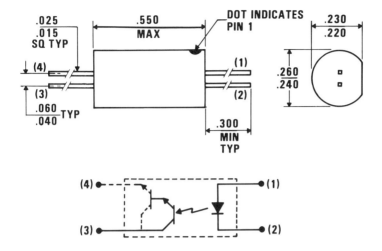

Figure 10.11 10 kVDC optocoupler. The physical separation of the emitter and sensor makes high isolation voltage possible. The design is commercially available with either a phototransistor or photodarlington output.

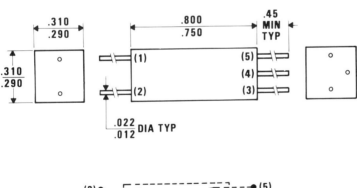

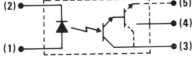

Figure 10.12 15 kVDC optocoupler. This particular design incorporates hermetic TO-46 and TO-18 discrete components. The military and high reliability electronics (e.g., aircraft) market is served by these products.

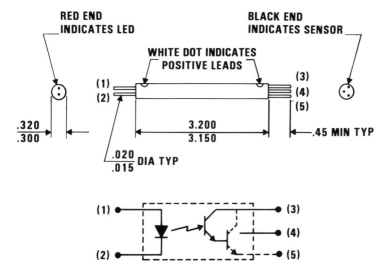

Figure 10.13 50 kVDC optocoupler. Increasing the physical separation between IRED and phototransistor makes very high voltage isolation possible. An internal high dielectric light pipe is necessary to transmit the infrared energy from IRED to sensor.

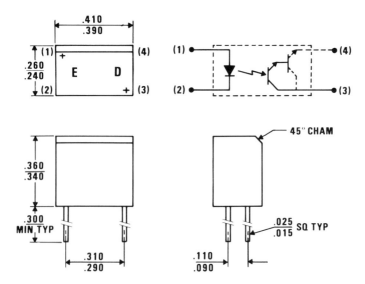

Figure 10.14 6 kVDC optocoupler. This design utilizes the lateral discrete devices shown earlier. The feature of easy PC board mounting is provided while relatively high isolation voltage is maintained.

10. Electrical Considerations

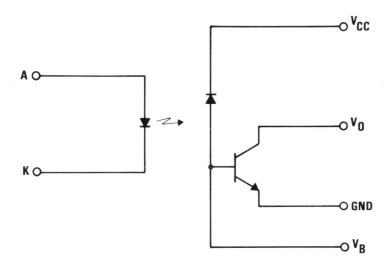

Figure 10.15   Electrical schematic of a simple optocoupler utilizing photo IC. High DC current transfer ratios are possible using a photodiode coupled into an output transistor. The 6N135 and 6N136 are electrically selected from lots manufactured using this design.

When the load resistance is kept in the 100 to 200 ohm range, the frequency response can extend in excess of 1 MHz. This compares to 25 KHz in a conventional phototransistor coupler.

A similar unit with higher gain characteristics but lower frequency response is the 6N138 and 6N139. The gain is increased from 7% to 300 percent while the data rate is reduced from one megabit/sec to 300 kilobits/sec. Figure 10.16 shows the electrical schematic of this optocoupler.

Another series of high technology couplers incorporate a Schmitt trigger in the photo IC output and are available in four different output options. Figure 10.17 shows the electrical schematic for these four options.

The 6N137 is a high speed (10 megabits/sec) optocoupler. The LED utilized is a visible red unit capable of these data rates. The solution grown epitaxial GaAs and GaAlAs IREDs are not fast enough. Figure 10.18 shows the schematic for this coupler.

These units are also available in other options such as packages, speed up resistors, duals, etc. Different types of these high technology couplers will become available as the speed capability is improved. These units are generally used in the telecommunication and the computer peripheral industries.

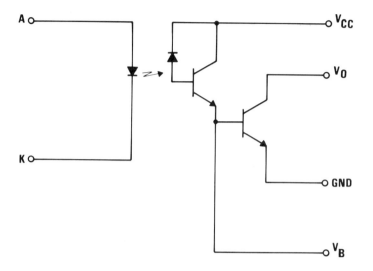

Figure 10.16 Electrical schematic of a darlington output optocoupler utilizing photo ICs. The photodiode output is amplified by a Darlington pair to increase CTR at the cost of operating speed. Note that an eight pin package must be used with this and the 6N135/6N136 designs.

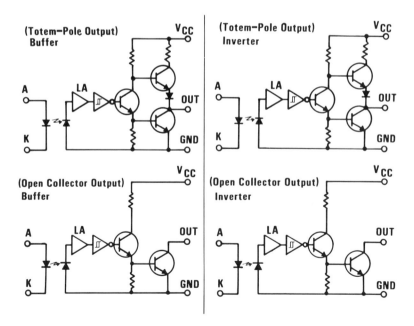

10. Electrical Considerations

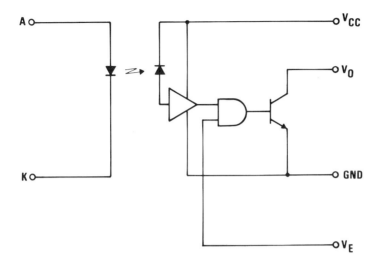

Figure 10.18 Electrical schematic of a high-speed coupler utilizing a visible LED and a photo IC. The LED used in the 6N137 design must be visible red in order to reach the switching speeds required by the design. Even faster speeds will soon be possible through the use of advanced materials (e.g., quaternaries).

Figure 10.17 Electrical schematic of a photo IC coupler with four output options. Logic couplers designed to drive TTL directly are manufactured using the same photo ICs that make high speed, digital output discrete photosensors. Data rates of 250 kbaud are easily transmitted with these designs. Packaging used in the popular six pin PDIP.

11

Mechanical Considerations

11.1 FABRICATION TECHNIQUES

The PDIP or plastic dual-in-line package constitutes approximately two-thirds of the sales dollars on couplers. Construction techniques differ significantly from one manufacturer to another. Figures 11.1 through 11.6 show the six most common construction techniques.

The sensor and IRED are mounted on two separate lead frames that have been formed to provide an offset or separation between the components. After mounting and bonding, the lead frames are spot welded together, the silicone gel light pipe and bond protector are added, and the outer case is molded around the unit. The standoff voltage is limited by the breakdown or "punch through" voltage along the silicone gel/molded plastic interface.

The construction technique of the double-molded PDIP is identical to the single-molded version except there are two molding operations. The inner mold material is a different type of plastic to offer better standoff voltage characteristics. Since its bond to the lead frame is not as good the moisture resistance characteristics are significantly degraded. As a result, the second coat is added to improve the seal quality and improve the moisture resistance. This approach is identical to the single-molded approach, except that a clear plastic membrane is placed between the sensor and IRED. This membrane improves the standoff voltage characteristics. However, it can damage the bond wires that are used to contact the emitter and base of the phototransistor and the anode of the IRED.

Again, this approach is quite similar to the single molded approach except a piece of glass is inserted between the sensor and IRED. This improves either the current transfer ratio for a given standoff voltage or improves the standoff for a given current transfer ratio. The four techniques described above are basically quite similar.

This technique utilizes the side emitting energy of the IRED as well as the energy from the top of the IRED carried through the silicone gel light pipe. A variation of this technique uses a molded plastic cup over the sensor and IRED. This smoothes the silicone gel, which in turn improves the energy transmission and current transfer ratio and makes it more consistent.

11. Mechanical Considerations

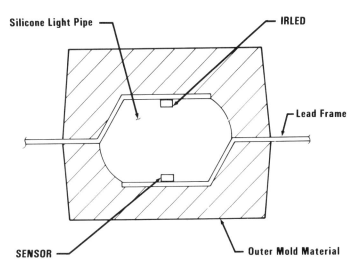

Figure 11.1 Single-molded PDIP. The simplest design utilizes silicone infrared transmission gel surrounded by transfer molded plastic.

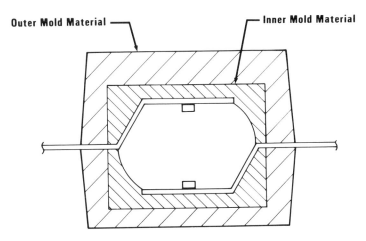

Figure 11.2 Double-molded PDIP. Increased reliability is obtained by using an inner mold of high dielectric plastic and an outer mold for secure lead support and higher moisture resistance.

Plastic Film

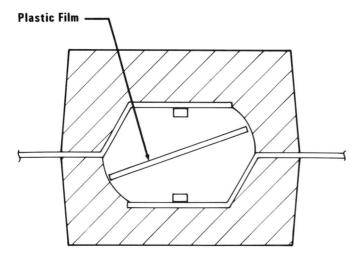

Figure 11.3 Single-molded PDIP with plastic isolation membrane. A further improvement in isolation voltage is possible by adding a high dielectric plastic film between emitter and sensor.

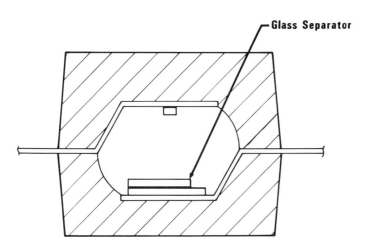

Figure 11.4 Sandwich construction of molded PDIP. A glass separator may substitute for the film shown in Figure 11.3, with an equivalent improvement in isolation voltage.

11. Mechanical Considerations 187

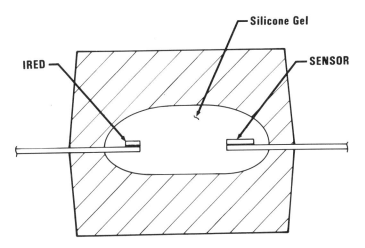

Figure 11.5 Single plane mounting construction for molded PDIP. The "side-by-side" or single plane mounting configuration affords higher isolation voltages than the "over/under" style illustrated in the previous four examples.

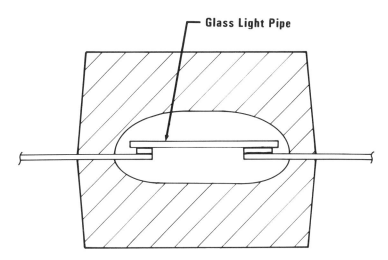

Figure 11.6 Single plane mounting configuration with light pipe for molded PDIP. The addition of a glass light pipe to the single plane configuration improves CTR while maintaining the isolation voltage advantage of this design.

Table 11.1   Techniques of Manufacturing

Technique	Advantages	Disadvantages
Figure 11.1	Good moisture resistance Temperature cycle Good CTR consistency	Lower isolation or CTR
Figure 11.2	Higher isolation or CTR	Moisture resistance Temp cycle Complications processing
Figure 11.3	Isolation Moisture resistance Temperature cycle CTR	Costly Lower yields
Figure 11.4	CTR Temperature cycle	Lower isolation Costly
Figure 11.5	Constant high isolation Low cost	Low and inconsistent CTR
Figure 11.6	Consistent CTR High CTR High isolation	Higher cost

This technique is similar to Figure 11.5, except that a glass plate is placed over the sensor and IRED. This glass acts as a light pipe that improves the current transfer ratio and makes it very consistent from one part to the next. Table 11.1 ranks these techniques by cost to manufacture and outlines the relative advantages and disadvantages.

11.2   OTHER MECHANICAL CONSIDERATIONS

The PDIP coupler is usually rated for a maximum storage temperature of 150°C. The plastic utilized in the unit has superior temperature characteristics when compared to the plastics utilized for discrete IREDs and photosensors. The pins as they egress from the plastic are 0.033 in. × 0.020 in. The portion that goes into the solder holes on a printed circuit board narrows down to a 0.020 in. × 0.020 in. size. This portion is usually covered with solder after mounting, and good heat transfer is provided as a result. Their resistance to solvents is the same as the transfer molded plastics covered in Part II.

11. Mechanical Considerations 189

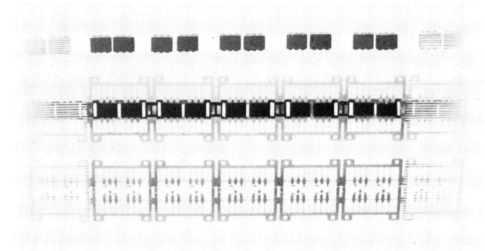

Figure 11.7 Photographs of varying process steps in PDIP manufacture. After the basic manufacture is completed, each part must be thoroughly electrically tested for many different parameters. Isolation voltage and CTR testing are the most critical.

The photographs shown in Figure 11.7 show the typical 16 unit lead frame after welding and silicone gel addition, after molding and prior to deflashing, and after clip and form.

V
OPEN AIR AND FIBER OPTIC COMMUNICATION

12
Fiber Optic Communication

12.1 BASIC THEORY

Fiber optics is an emerging technology. It basically consists of a transmitting section that includes a signal-processor circuit for modulating or driving an IRED or laser, the IRED or laser, and a means of coupling the energy into a plastic or glass fiber for carrying the information. The plastic or glass fiber will be of a length dictated by the application with repeater stations if the fiber is long enough. The receiving section would include a method of coupling the energy from the plastic or glass fiber to a photosensor to detect the signal, and a signal processor to process the output of the photosensor. Figure 12.1 shows the cross section of a fiber optic cable.

The energy is injected into the N_1 material. The N_1 material acts as a wave guide. Snell's law (as previously discussed on IREDs and assemblies) is followed utilizing the core material as the N_1 wave guide with the cladding acting as the N_2 material. There are two basic types of single mode fibers that are used for transmission of the optical information or data: the step-index fiber (most popular) and the graded core fiber. The step-index has an abrupt change of refractive index at the N_1/N_2 (core/cladding) interface. The graded core fiber has a gradual or graded change of refractive index from the N_1 core to the N_2 cladding.

Plastic fibers are utilized for some short distance transmissions due to their low cost. Glass fibers, having a higher cost, are more efficient and are primarily used for long-distance transmissions. Their efficiency improves as the wavelengths get longer. The two most popular wavelengths are 820 nm (0.82 µm) and 1300 nm (1.3 µm). The 820 nm is used with IREDs and both plastic and glass fibers. The 1300 nm is primarily used with glass fibers. The 820 nm is used with a silicon photosensor at rates up to 40 MBd, while the 1300 nm operates with a laser source and GaAs receiver at much higher information rates (400 MBd). The plastic fiber is usually larger in cross section (approximately 1mm) while the glass fiber is 85/125 or 85/140 µm in diameter. There are other sizes of glass fibers used but the larger core diameter (85 µm) allows maximum energy to be coupled into the fiber. The glass fiber is usually specified with a numerical

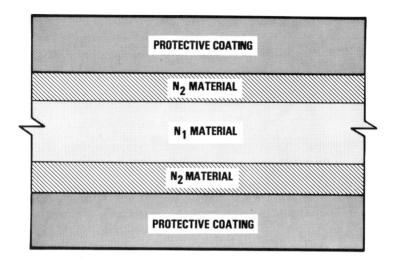

Figure 12.1 Cross-sectional drawing of a fiber optic cable. The difference in the refractive index between N_1 and N_2 establishes a critical angle large enough to reflect virtually all of the traveling infrared signal back toward the center of the fiber. Very little transmission loss occurs through the side of the fiber.

aperture (such as 0.29). This is the inverse sine of the half-angle of radiated energy the fiber will accept and propagate. Thus if a fiber is characterized by:

85/125 with NA of 0.29

The diameter of the optical carrier core is 85 μm, the diameter of the cladding is 125 μm, and the core will accept energy striking it at ±17°C from the longitudinal centerline.

The basic principle of fiber optic communication is quite simple. Coded pulses of energy (either analog or digital) are coupled into a fiber, propagate along it, and are detected by a sensor.

12.2   BACKGROUND

Fiber optic transmission of data offers a number of significant user advantages over wire or radio transmission.

Higher Information Density

The high bandwidth distance product allows a single fiber to carry more information (400 MBd/20,000 Hz = 10,000 conversations).

12. Fiber Optic Communication

EMI (Electromagnetic inference) and RMI (Radiofrequency Inference)

The plastic or glass fibers are immune to external electromagnetic interference. Lightning, solar activity, automotive ignition, nearby power generators, and spurious interference from radio frequency sources such as ham operators have no effect on the information carried within the fiber. This becomes extremely important in data and video communications as well as selected control applications.

Small Size and Lighter Weight

A typical optical cable weighs one-third to one-tenth the weight of a copper cable of equal diameter. When we consider both the weight and higher information density factors, the fiber optic appeal becomes very marked. This factor is of particular interest to the aircraft, automotive, and naval industries.

Fire Hazard Prevention/Short-circuit Protection

Optical fiber does not carry current, hence cannot generate heat or sparks. This significantly reduces the fire hazard. The chemical, petroleum, and other safety conscious industries are particularly interested in this property. This also allows installation without conduits, thereby saving cost or allowing use in existing wire troughs, conduits, elevator shafts, or through ventilation ducts.

Security

Fiber-optic links are secure, because they cannot be tapped without the knowledge of the user. This is especially important in military systems and many commercial systems.

Electrical Isolation

In one sense, fiber optic links are optocouplers, as discussed in Part IV. Commercial optocouplers are limited to approximately 10 MBd at present. The higher frequency response of the fiber-optic links gives the optocoupler isolation at computer speeds. Ground-loop problems are eliminated. The trend to expand computer control of heavy electrical equipment in large manufacturing facilities dictates the use of fiber optic transmission.

12.3 FIBER OPTIC TECHNICAL BARRIERS

The trend toward fiber optic transmission of data has been hampered by some major drawbacks.

LACK OF STANDARDIZATION: In the early stages of an emerging technology, each supplier presents their own ideas on how a component should be packaged and specified for performance. This generally leads to a lack of compatibility between designs. As the use of these products builds up, the user generally will select the parts best suited for the application and, by virtue of the quantity involved, will demand and receive second sourcing to those same specifications. Fiber optic transmission has matured to the point that standardization is starting to occur. However, this lack of standardization in the past has slowed its growth.

TERMINATION DIFFICULTIES: Existing connector designs generally rely on polishing operations and epoxy adhesives to make the connection of fiber to fiber. This problem is accentuated by the low volume demand and the degree of difficulty in obtaining a low loss joint or connection between the two glass fiber ends. An easy to use, low cost connection or splicing method must be developed to hasten the widespread acceptance of fiber optics.

ONE-WAY COMMUNCATIONS: Utilization of a fiber requires a transmitter at one end and a receiver at the other end. As a result, most of the present applications are limited to one way (simplex) communications. Half-duplex capability can be achieved by the utilization of couplers on each end of a single fiber. This allows the data to be transferred in one direction or the other but not simultaneously (hence, half-duplex). In order to have full-duplex capability or simultaneous bidirectional communication, it is necessary to use two links in parallel.

FRAGILITY: Cable durability continues to be improved. However, care must still be exercised to protect the fiber from crush or shear stresses. These can result in breakage and disruption of data communications. Compounded by the lack of a cost-effective splicing method, this drawback continues to impede the market acceptance of fiber optic systems.

13

Open Air Communication

13.1 BASIC THEORY

Open air communication is just what the title implies. It is similar to the transmission of data by fiber optic transmission except the energy is not contained within a fiber. Since the transmitter and receiver are generally located several feet apart, the energy transmitted follows a square law falloff. For relatively short distances (up to 30 ft), commercial IREDs are used without a separate focusing lens. For longer distances (up to two miles) a separate focusing lens is necessary.

The IRED is pulsed at a low duty cycle and a high current to provide the transmitting energy. A photosensor, with associated high gain amplifier, is used to detect the transmitted data. The pulsing of the IRED and the receiving photosensor will be discussed in additional detail in Chapter 14.

Remote control of TV sets was one of the first applications of open air communication. Three GaAs IREDs with an emission angle of 50° between half power points were placed in series and used as a transmitter. A photodiode and amplifier were located in the TV set. The hand-held battery operated transmitter would send coded data to the receiving photodiode. Some of the energy would be directly received by the photodiode. More of the energy would simply illuminate the room by bouncing from walls and ceiling to the receiving photodiode. This concept had significant advantages over the ultrasonic systems used before. The transmitted data was much more impervious to spurious noise and was capable of transmitting and receiving more complex data. Figure 13.1 shows some of the typical encoding schemes used to transmit information.

Longer distance transmission utilizes a focusing lens but requires more precise alignment between the transmitter and the receiver. Molded acrylic lenses can be fabricated that will project a 10 to 15 ft circle at 500 ft. A GaAlAs IRED in the T-1 3/4 package gives a typical power output into a 0.250 in. diameter area located 1.130 in. from the lens tip of 120 mW/cm^2 when driven at 3 amps. If the focusing lens captured this power and dispersed it into a 10 ft diameter circle at 500 ft, the intensity would drop to approximately 240 μW/cm^2. A

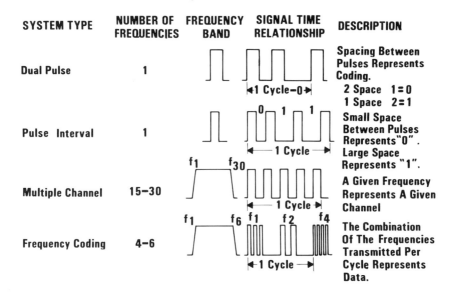

Figure 13.1 Typical IRED encoding methods. Pulse coding can have many variations, far advanced from the original Morse code, but still using the same principle.

1.5 in. convex lens would illuminate a 0.100 in. × 0.100 in. photodiode with approximately 171 μwatts of IR energy. The photodiode would sink approximately 5 μA of photo current. A building to building communication link or a perimeter surveillance system could be easily built using this approach. A laser could be substituted for the IRED and lensing system. The benefit of the lensing system can be quickly seen if the same calculation is made where distance is the variable for the 50° wide angle T-1 3/4 and the 16° narrow angle T-1 3/4. Table 13.1 shows these distances. A 4X magnifying lens and photodiode is used with a 0.100 in. × 0.100 in. photosensitive area.

Table 13.1 Comparison of System Without Magnifying Lens

16° T-1 3/4 @ 1.4 A	approx. 30 in. separation
50° T-1 3/4 @ 1.5 A	approx. 6. in. separation

13. Open Air Communication   199

A much more sensitive receiver system is required with the broad band broadcasting system such as is used in remote control of TV and CATV systems. The receiving system is discussed in more detail in Chapter 14.

13.2 BACKGROUND

The principles outlined in Section 13.1 offer exciting opportunities for infrared data communication systems. These may be divided into three basic categories and will be discussed with the assumptions that:

1. The IRED is driven at approximately 3 A peak current.
2. The receiver will detect 1×10^{-9} A pulses.

The first category is a hand-held transmitter that will operate effectively up to 30 ft. The target area or accuracy of pointing would be an area 28 ft in diameter. A standard T-1 3/4 GaAlAs IRED with a beam angle of 50° between half power points is the transmitter. The receiver is a photo diode (0.108 in. × 0.108 in.) mounted in a TO-5 package. The transmitter and receiver diagram are shown in Figure 13.2.

The emitter driver, IRED, photodiode, and amplifier could be built up in 1000 lot quantities for approximately $5.00. Chapter 14 will go into more detail on this. This type would find wide application in:

1. TV and CATV remote control
2. TV games remote control
3. TV antenna remote control
4. Stereo system remote control
5. Appliance remote control
6. Communication between keyboard and CRT
7. Communication between "mouse" and CRT
8. Interrogation of vending machine
   a. Service man—stock levels
   b. Supervisor—service control
9. Automotive interior communication—window control
10. Automotive exterior communication—door lock control
11. Waiter efficiency communication to bar or kitchen

The second category is an aimed hand-held transmitter and/or receiver. The aiming would require hitting a 100 ft diameter target. The distance would be controlled by the focusing lens that would be added in front of both the transmitter and receiver. Table 13.2 shows some of the angular divergence of lens versus distance where a 100 ft diameter circle is the consistent target.

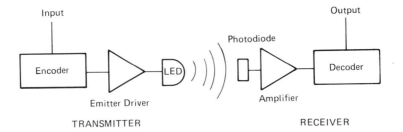

Figure 13.2 Simplified schematic of an infrared data communication system. Every infrared television remote control follows this basic schematic. Each of the components is readily available from various manufacturers as a standard product.

The only changes to the system would be the focusing lens added in front of the IRED and photodiode. If these were molded of clear acrylic plastic, the cost addition would be slight. This would open up a new variety of applications such as:

1. Garage door openers
2. Automotive luxury options:
   a. Remote engine starting
   b. Remote turn on or heater/air conditioning
   c. Remote raising or lowering of windows
3. Beam interrupt security systems
4. Remote control of mechanical systems:
   a. TV cameras
   b. Pilotless aircraft
   c. Mechanical toys
   d. Commercial sprinkler systems
   e. On/off valves
5. Remote interrogation:
   a. Meter reading

The third category is quite similar to the hand-aimed transmitter and receiver. However, the target size would be reduced to either improve the accuracy and the range, or to reduce the possibility of detection by outside systems. A decrease to 0.25° in the diverging angle of the focusing system would allow the development of a vehicle speed measuring system that would detect speeds at up to a 1/2 mile range in the reflective mode. Using this same type system in the transmissive mode, data could be transmitted up to two miles.

A low-cost communication link becomes very practical, provided that system limitations are not exceeded. The line of sight between the transmitter and the receiver must be free of IR blocking material.

13. Open Air Communication

Table 13.2 Angular Divergence of Focusing Lens versus Distance to Target

Diverging angle	Distance to target
27°	100 ft
14°	200 ft
7°	400 ft
3.5°	800 ft

The data rates must be low enough to stay within the average current limitations of the IRED. Higher data rates may be obtained by using a laser as the transmitter at a proportionately higher cost.

14

Pulse Operation

14.1 THE IRED

The advantage of running the IRED at high currents becomes obvious when the output versus IRED current is examined. Figure 14.1 shows this relationship on a broad and narrow beam T-1 3/4 units. The power shown is apertured power measured on a photovoltaic detector 0.250 in. in diameter and located 0.500 in. from the lens side of the mounting flange on the broad beam unit and 1.429 in. from the lens side of the mounting flange on the narrow beam unit.

At 3.0 amps forward current, the broad beam unit will typically emit 475 mW/cm^2, while the narrow beam unit will typically emit 114 mW/cm^2. In using these units in the pulse mode, there are three variables to take into consideration (the height of the pulse, the width of the pulse, and the repetition rate of the pulse). The average current (the pulse height times the duty cycle) is related to the maximum DC current allowed. However, the unit is receiving a cycling stress (pulse current causing internal heating and cooling) in addition to the average current. For conservative designs, the maximum average current (pulsed) is limited to two-thirds the maximum DC current. These results indicate the degradation rates of the IREDs under these two conditions are approximately the same (again a conservative approach). Figure 14.2 shows the degradation rates at an average current of 1 mA (I_{peak} = 1 A, PW = 100µs, 10PPS) and a DC current of 20 mA. The degradation rate at 10,000 hours for the 1 mA average current is 2.5% while the degradation rate for 20 mA DC is 7.5%. This would equate to the one-third derating factor. Refer back to the section on reliability in Chapter 3 for the total use of these curves.

The curve shown in Figure 14.3 shows the maximum peak pulse current versus pulse width. Note the slope change after pulse widths exceed 25 µs. This is due to the internal heating caused by the power pulse. Note also at 300 ms that the curve becomes the maximum DC rating for the part.

The maximum peak current is obtained from two factors. The nonlinearity in output versus current density is caused by current

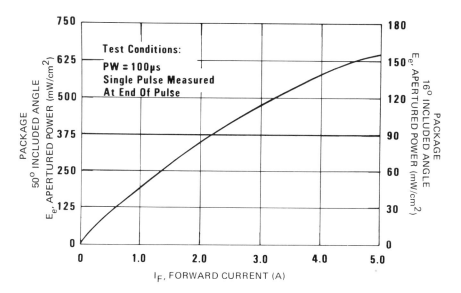

Figure 14.1 Apertured power versus forward current for T-1 3/4 GaAlAs IREDs. For high forward pulse currents, IRED output rises dramatically. The energy level is analogous to a flash bulb going off, although unseen by the unaided eye.

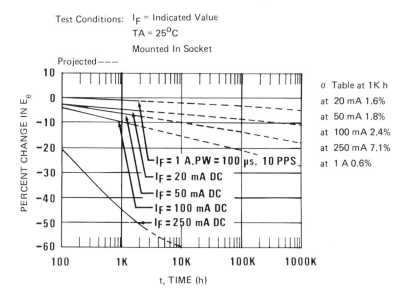

Figure 14.2 Percent change in power output versus time for T-1 3/4 GaAlAs IREDs. Pulsing adds the additional stress of the current cycle; however, empirical data demonstrates that carefully controlled high current pulse operation is no more harmful than low current continuous operation.

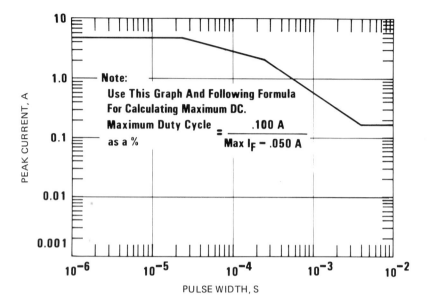

Figure 14.3 Maximum peak pulse current versus pulse width for T-1 3/4 GaAlAs IREDs. The "envelope" for safe operation is shown. Safe operation is defined as having an extremely low chance of catastrophic failure and acceptably low output degradation.

crowding within the chip. The physical limitation in the bond wire's ability to carry higher currents results in an overall maximum current higher than and independent of the average current.

The method of pulsing the IRED is rather straightforward. Typically, the battery or power supply charges a capacitor during the off cycle. Figure 14.4 shows a simple schematic utilizing a 9 V battery and assuming a 1 V drop across the switching transistor in the "on" mode.

The average I_F of the IREDs at a forward voltage of 4 V is 3.4 amps. As the battery ages to 7.5 V, the average I_F will drop to 2.25 amps. If the selected transistor switch has an "on" voltage less than 1.0 V (the usual case), then a current limiting resistor is used to control the current. Under the above conditions, the peak current pulse width would be a maximum of 80 μsec and the maximum duty cycle would be 3.3% (see Figure 14.3).

$$\text{Maximum duty cycle is a \%} = \frac{0.100 \text{ A}}{3.040 - 0.050 \text{ A}}$$

$$\text{Duty cycle} = 3.3\%$$

14. Pulse Operation

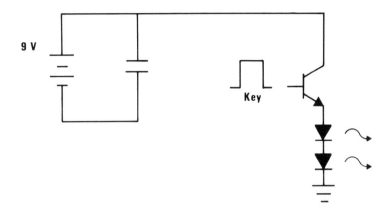

Figure 14.4 Simple schematic of pulse circuit for high current pulsing of two T-1 3/4 GaAlAs IREDs. Just as one might logically expect, the high current is provided by a charged capacitor with switching controlled by a transistor.

In order to demonstrate how to determine the usable power at the receiver, two calculations will be used as examples. The first system will utilize a focusing lens designed to focus the IRED energy within a 25 ft diameter circle. The assumption is made that the focusing lens will capture all of the energy within a 30° included angle from the IRED. The system's worst case of $I_F = 2.25$ A will be used. The effective power (from Figure 4.1) is 375 mW/cm^2 (50° included angle). The total energy within the 25 A diameter circle is equal to the amount of energy within the 0.250 in. diameter circle.

$$\text{Ratio} = \frac{r_1^2}{r_2^2} = \frac{(0.125/12)^2}{(12.5)^2} = \frac{0.0013}{156.25} = 8 \times 10^{-6}$$

$$\text{Energy} = 375 \text{ mW/cm}^2 \times 8 \times 10^{-6}$$

$$= 3000 \times 10^{-9} \text{ W/cm}^2$$

$$= 3 \text{ }\mu\text{W/cm}^2$$

This number will be used in Section 14.2 for a sample calculation. The distance between the transmitter and the receiver is now a variable to be considered in lens design. Table 14.1 shows some varying distances between the transmitter and receiver for the diverging angle of the lens.

Table 14.1 Distance versus Diverging Angle

Distance	Tan θ	θ (diverging angle)
1/16 mile	0.0379	2° 10 ft
1/8 mile	0.0189	1° 5 ft
1/4 mile	0.00946	32 ft

A larger diverging angle can be obtained by decreasing the target size. This would also increase the energy/unit area. The 3 µW/cm^2 would be quadrupled by lowering the target area from a 25 ft diameter circle to 12.5 ft diameter. A good two stage amplifier can detect 20×10^{-9} W/cm^2, which is 1/150 of the 3 mW/cm^2 used in the example.

The second system used for demonstrating sample calculations will utilize a narrow beam T-1 3/4 GaAlAs IRED without an external focusing lens. It will assume 90 mW/cm^2 (Figure 14.1) at a distance of 1.129 in. from the lens tip. If the assumption is made that 1×10^{-9} W/cm^2 is detectable and that the energy fall-off follows an inverse square law relationship, then Table 14.2 shows the transmission distance. This would correspond to a distance of approximately 1000 ft, with the location accuracy of the sensor a 175-ft diameter circle. Converting to a flat window photocell without the 4X magnifying factor would reduce the range to approximately 500 ft, with a 86.5-ft diameter target area. Table 14.3 summarizes the same type of calculation utilizing the broad emitting angle T-1 3/4 GaAlAs IRED with and without the 4X magnifying lens on the photo sensor.

Table 14.2 Distance versus Power/Unit Area

Distance	Power/unit area
1.129 in.	90 mW/cm^2
Approx. 4 in.	5.6 mW/cm^2
Approx. 16 in.	350 µW/cm^2
Approx. 64 in.	22 µW/cm^2
Approx. 256 in.	1.4 µW/cm^2
Approx. 1024 in.	87.5 nW/cm^2
Approx. 4096 in.	5.5 nW/cm^2
Approx. 16384 in = 1362 ft	0.34 nW/cm^2

14. Pulse Operation

Table 14.3 Comparison of Distance versus Units

Unit	Target area 4X lens receiver	Target area 1X lens receiver	Distance
50° T-1 3/4	214 ft diameter		400 ft
16° T-1 3/4	175 ft diameter		1000 ft
50° T-1 3/4		107 ft diameter	200 ft
16° T-1 3/4		87.5 ft diameter	500 ft

14.2 THE PHOTOSENSOR

The photosensor usually consists of a photodiode and a high gain amplifier. Figure 14.5 shows the block diagram of a commercial PIN photodiode and a high gain amplifier.

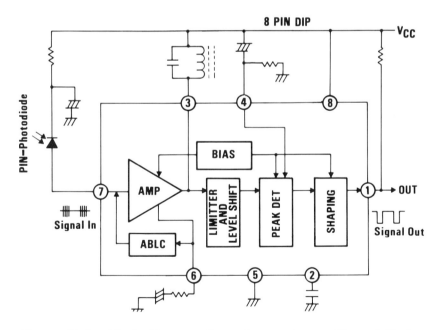

Figure 14.5 Block diagram of a pulse receiver. The relatively small current from the PIN diode must first be amplified prior to further signal processing. This function, plus signal processing circuitry, is available in a standard eight pin DIP.

Table 14.4 Pin Function for Pulse Receiver

Pin 1	Output
Pin 2	Internal capacitor—carrier wave through peak detector is integrated by this capacitor. The time constant is determined by this capacitor and the resistor connected between pin 1 and V_{CC}.
Pin 3	Tuning coil—Typically set for 40 kHz. This can be used to filter background noise. If input pulse is made up of a burst of 40 kHz pulses for the carrier, the background noise other than 40 kHz will be filtered.
Pin 4	The signal output from the tuning coil terminal (pin 3) is detected by the peak detector circuit. The detecting level depends on the input signal strength so the noise is suppressed. The time constant of the peak hold is changed by the RC network connected to pin 4.
Pin 5	Ground.
Pin 6	Bypass capacitor—The ABLC (automatic bins level control) slowly changing energy signals (ambient light such as sunlight) from effecting the receiver. It is effectively a low frequency roll off network. The amplifier has a gain of load with a resistor value of 22 ohms.
Pin 7	Input—The input impedance is typically 10K.
Pin 8	V_{CC}—The V_{CC} can vary between 6.0 to 14.4 V.

If an input of 3 $\Omega W/cm^2$ (from Section 14.1) is assumed with a PIN photodiode (0.108 in. × 0.108 in.) that has no magnifying lens, the minimum photo current should be 30 nA. The minimum input impedance of the amplifier is 40K. This would create an input voltage of 1.0 mV. The amplifier is designed to function with 100 μV at the input. A ten-to-one safety factor is present. A pin by pin function is listed in Table 14.4.

Figure 14.6 shows the amplifier gain versus the resistor value tied to pin six. Figure 14.7 shows the output voltage peak to peak at pin 1 versus the resistor value between pin 4 and ground.

There are other commercial amplifiers available to perform the same function. The ABLC feature operates until the ambient light

14. Pulse Operation

level becomes strong enough to saturate the amplifier. The 40 kHz carrier is a carry over from the ultrasonic transmitter receiver used in early remote control systems for TVs. Figure 14.8 shows a pulse train from Figure 13.1 where the 40 kHz carrier makes up each individual pulse.

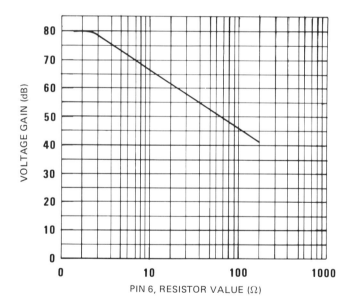

Figure 14.6 Amplifier gain versus pin 6 resistor value. Sensitivity related to the amplifier gain (Figure 14.5) may be adjusted by changing the value of the resistor at pin 6.

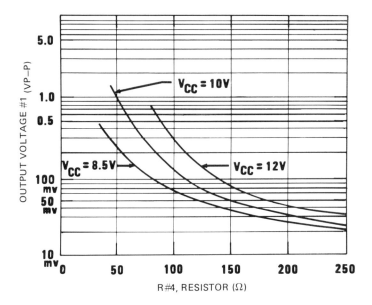

Figure 14.7 Sensitivity of peak detector characteristics. The pin 4 resistor, also connecting V_{CC} to ground through its resistance, acts as a voltage divider to ultimately control the output at pin 1.

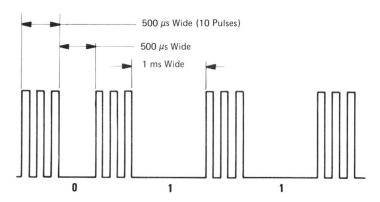

Figure 14.8 Pulse train utilizing a 40 kHz carrier. The choice of 40 kHz is a carry-over from earlier ultrasonic remote control units (just above audio frequency, but low enough to minimize RFI). Other frequencies may be received by adjustment of the LC filter on pin 3 (Figure 14.5).

VI
OPTOELECTRONICS APPLICATIONS

15
Driving the IRED

15.1 INTRODUCTION

Pulsing the IRED was discussed in Chapter 14. This chapter will deal with the method of bias or the control circuitry for the IRED in the DC mode. Figure 15.1 shows a simple schematic for a constant current supply to the IRED.

Calculation of the resistor value is a simple application of Ohm's law. If a voltage drop of 1.6 V across the IRED is assumed for 20 mA forward diode current, then:

$$R = E/I = \frac{5 - 1.6}{0.02} = \frac{3.4}{0.02} = 170 \ \Omega$$

The 20 mA I_F and the 3.6 volts allow the calculation of wattage size:

$$P = EI = 3.6 \times 0.02 = 72 \text{ milliwatts}$$

The resistor value is usually chosen to allow a minimum current flow of 20 mA. Thus the selection would allow for resistor tolerance and be the standard value below 170 Ω including tolerance. The wattage would be the first standard value above 72 milliwatts.

15.2 VARIABLES

The forward voltage drop of an IRED will vary from unit to unit. A typical GaAs IRED is specified at 1.6 V maximum forward voltage at a forward current of 20 mA at 25°C. The distribution of parts will be centered at 1.2 V and range from 1.16 to 1.6 V. A recalculation of forward current utilizing the 170 Ω resistor and the circuit shown in Figure 15.1 shows a variation of 20 mA to 22.6 mA. The nominal current for the 1.2 V will be 22.3 mA.

The power supply of five volts will also have some variation. This variation will cause a variation in the forward current. The resistor tolerance would also cause a variation in the forward current.

Temperature shifts will cause the forward voltage drop and the output energy to shift. These are somewhat compensating. Figure 15.2 shows both the change in V_F and P_O for a GaAs IRED.

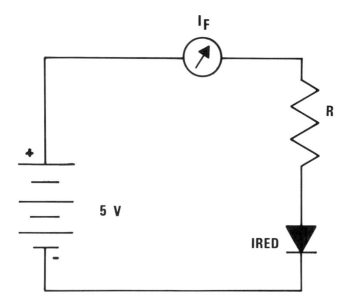

Figure 15.1 Bias circuitry for IRED. For continuous operation, a voltage source and current limiting resistor are the only components needed to drive the IRED.

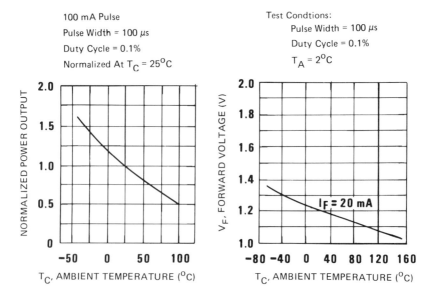

Figure 15.2 Changes in forward voltage and power output versus temperature. Output drops as temperature increases, but forward voltage also drops (allowing current to increase if voltage is left constant). In actual practice, these two changes partially offset one another, although the dominant effect is the output drop.

15. Driving the IRED

Table 15.1

Internal heating	20%
5% power supply	4%
IRED degradation	10%
Net loss	34%

The typical value of forward voltage will decrease from 1.2 V to 1.1 V as the temperature is increased from 25° C to 100° C. This would allow the current through the IRED to increase from 22.3 mA to 22.9 mA. This 2.7% increase would partially compensate for the 50% decrease in output power due to the temperature increase.

The heating effect within the IRED caused by the current through it and the voltage drop across it will normally cause the IRED to run hotter than its mating photosensor. At 20 mA forward current and 1.6 V, the dissipated power will be 32 mW, with a derating of 1.33 mW/°C. The junction temperature would be approximately 49° C.

$$T_j = 25° C + \frac{32 \text{ mW}}{1.33 \text{ mW/°C}} = 49° C$$

This increase in junction temperature would cause a decrease in output of approximately 30%.

The effective IRED current is reduced by 34%. By reducing the IRED current from 20 mA to 13.2 mA in an in-line test of the system, the functionality could be guaranteed from these variables. If other variables needed to be covered, a simple addition to the table and subsequent calculation could encompass them. These are shown in Table 15.1.

15.3 LINEAR OPERATION

At the present time, the characterization of IREDs provides inadequate data for linear operation design. The devices are usually characterized and specified for digital design using opaque beam interrupt mechanisms to provide the switching from the "on" to "off" state. Figure 15.3 shows the curves of output power and forward voltage versus current for a typical GaAs IRED.

If the IRED was biased for a steady state forward current of 25 mA then a reasonably linear AC coupling could be obtained with a ±12 mA swing. Since heating would become a factor at the higher currents, the IRED maximum current should be as low as possible. The IRED becomes nonlinear at lower currents so the minimum current should be kept above the 1 to 2 mA region. The reader should now review Part I and the points made will have more continuity.

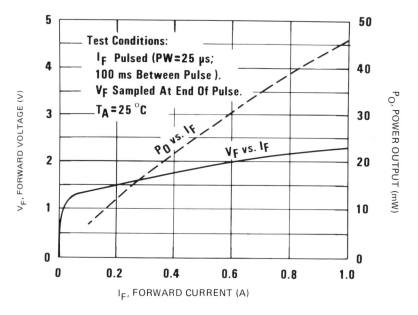

Figure 15.3 Output power and forward voltage versus current. Note that the diode does not begin to conduct until the forward voltage passes almost 1 V. Beyond this point, both curves are reasonably linear; however, secondary effects, such as junction heating, can dramatically affect output.

16

Interfacing to the Photosensor

16.1 THE PHOTODIODE

The PN or PIN photodiode is usually utilized in applications where speed of response or linearity of output over a wide range of levels of input is required. The photodiode can switch in a few nanoseconds and respond linearly to input levels over nine orders of magnitude. The output sink current varies from the high picoampere to the low milliampere range. This output signal is normally amplified. The input impedance of the amplifier used is normally low to minimize the RC (R-input impedance, C-diode capacitance) time constant of the system. Figure 16.1 shows a typical schematic of a PIN photodiode and amplifier.

The amplifier shown is a transverse impedance amplifier with the value of R chosen to set the gain. Typically, the higher the gain, the lower the frequency response and the higher the input impedance. Operational amplifiers are also widely used for this type of application.

16.2 THE PHOTOTRANSISTOR AND PHOTODARLINGTON

The discussion in Part II on the transistor used as the photosensor pointed out that the switching characteristics of the phototransistor are inferior to the conventional small signal transistor. This is due to the enlarged base area of the phototransistor that increases the collector to base capacitance. Significant variables also affecting the switching time are the depths of the junctions, which are optimized for photon recombination, and the spacing or geometry of the transistor, which is usually larger on the phototransistor.

The phototransistor is normally used as a switch. When sinking current, it is considered "on" and when not sinking current, it is considered "off." Figure 16.2 shows a simple schematic.

When energy is impinging on the base region, the transistor is in the conducting mode. The voltage drop across the load is high being $V_{CC} - V_{CESAT}$. When the energy is blocked or removed, the only conduction current is the leakage current. The voltage

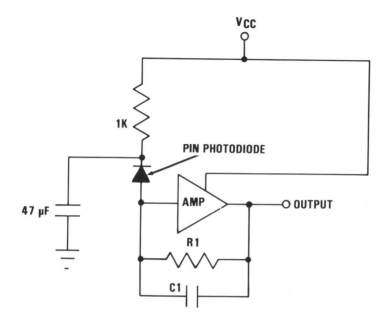

Figure 16.1 Simple pin diode and amplifier schematic. A low impedance amplifier is normally used to maximize the speed of the PIN photodiode.

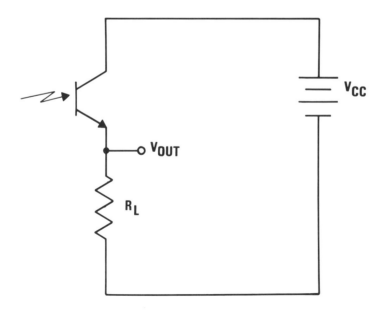

16. Interfacing to the Photosensor

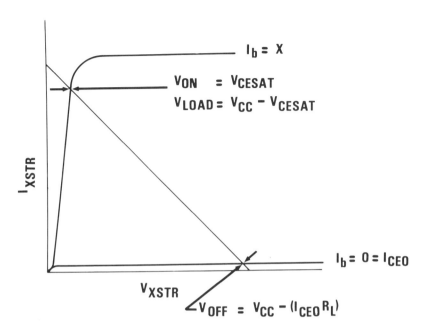

Figure 16.3 V_{ON} and V_{OFF} for a phototransistor operating as a switch. The transistor voltage drop is at its maximum when "off" (nonconducting dark state) and becomes minimal when turned to the "on" state.

drop across the load approaches zero. Figure 16.3 shows the voltage versus current for these two states.

The two conditions of the switch are thus specified. The voltage drop across the load must be large enough to actuate the switch in the "on" condition and must be small enough in the "off" condition to cause a switch state change. Figure 16.4 shows the circuit when driving a TTL gate.

When the transistor is "on," the V_{OUT} must be below 0.8 V and the sink current must be $\geqslant 1.6$ mA. When the transistor is "off,"

Figure 16.2 Simple schematic of the phototransistor operating as a switch. Perhaps the most simple circuit possible is to have a voltage source and load resistor (emitter to ground) in series with the phototransistor. Switching time is dependent upon the value of R_L and the speed of the phototransistor.

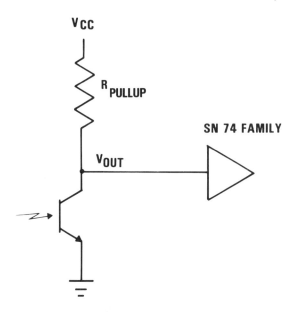

Figure 16.4 Driving a transistor logic (TTL or T^2L) gate. In this case, the "off" state creates a voltage into the gate. The "on" (transistor conducting) state drops the gate voltage to less than 0.8 V.

the V_{OUT} must be $\geqslant 2.0$ V. This must hold true over the operating temperature range of the system and not exceed 0 to 70° C. The output current sink capability of the phototransistor may not meet the 1.6 mA minimum at a maximum voltage drop of 0.8 V. Figure 16.5 shows this condition.

An increase in I_b or the energy impinging on the base of the phototransistor can solve the problem. If this is not possible, then the SN74 series unit could be replaced by the low power versions of the SN74 series. The sink current applied must be $\geqslant 0.180$ mA. Another solution is to use an intermediate amplification stage, as shown in Figure 16.6.

The addition of the 2N2222 lowers the output requirements of the phototransistor. R_1 is chosen as a current limit resistor for the phototransistor and must be large enough to provide adequate drive to turn the 2N2222 "on." R_2 is the pull-up resistor that provides a path to supply the I_{CEO} of the 2N2222 when V_{OUT} is high.

The photodarlington is similar in output characteristics to the phototransistors with two exceptions. The $V_{CE(SAT)}$ characteristics

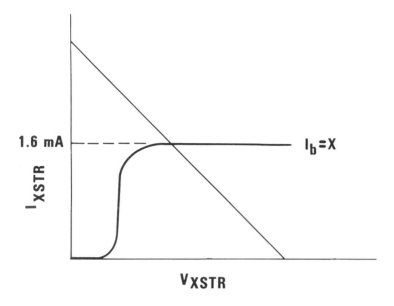

Figure 16.5 $I_{OUT} < 1.5$ mA @ $V_{OUT} = 0.8$. In this case, the transistor will not sink sufficient current to pull the gate voltage to its low state. Either a low current TTL gate or additional amplification stage is required.

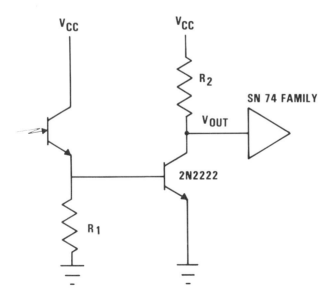

Figure 16.6   Improving the sink current with an intermediate gain stage. An intermediate amplification stage may be as simple as the addition of a standard switching transistor circuit (e.g., 2N2222).

of the photodarlington are higher than those of the phototransistor by one forward voltage drop of approximately 0.7 V. The cascaded structure of a transistor feeding a second transistor causes this problem. The switching times of a photodarlington are usually more than an order of magnitude higher than a phototransistor with the same effective load resistor. The added gain of the photodarlington, however, can be useful in certain application areas. This gain difference is normally a factor of slightly over 1 to greater than 250.

If the application requires the extreme upper end of sensitivity of a phototransistor (h_{FE} of 1000 to 1200) the cost may be prohibitive or the distribution may yield too few units to satisfy the volume. Photodarlingtons with h_{FE}s in the 1000 to 3000 range are relatively easy to make and thus are far more suitable than high gain phototransistors. The photodarlington is built with a common collector region. The simplified gain is nothing other than the square of the transistor gain. A photodarlington with a gain of 900 (30 × 30) is much easier to fabricate than a phototransistor with a gain of 900. Figure 16.7 shows the schematic for both a phototransistor and a photodarlington.

The second type of application involves the need for higher output current with a given energy level input. If the energy level is

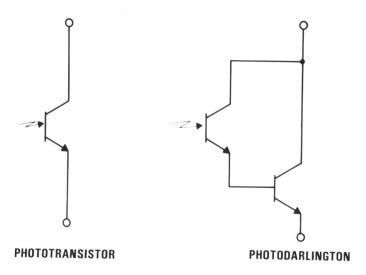

PHOTOTRANSISTOR   PHOTODARLINGTON

Figure 16.7 Schematic of a phototransistor and a photodarlington. The gain advantage of the photodarlington over the phototransistor may be as much as a factor of 250. However, speed is usually slower by an order of magnitude.

16. Interfacing to the Photosensor

in the microwatts range, the high gain photodarlington allows adequate gain for a usable output signal. This can be useful in applications such as a "touchscreen," where the optical switches have separation of transmitter and receiver that exceeds the normal 0.050 in. to 0.500 in. If the input energy level is in the high microwatt, low milliwatt range, the high gain photodarlington may be used to drive relays directly. There is clearly a place in the optosensor line for applications of the low cost phototransistor and photodarlington.

16.3 THE PHOTOINTEGRATED CIRCUIT

The photo IC is the easiest unit to interface to because it is normally specified to be compatible with logic circuitry and is usually guaranteed to operate under specified conditions over a broad temperature range. It offers the advantages of the photodiode input and conditioning or logic circuitry output. The only major disadvantage is the premium price. The future trend will be toward availability of more special function options, decreasing price, and erosion of discrete photosensor applications as these units become more widely used.

17

Computer Peripheral and Business Equipment Applications

17.1 WINCHESTER DRIVE

An optical encoder is normally used to identify the different tracks on the disk platter. In addition, an interruptive optical sensor is used to identify track zero which is the home position or "write" start. Winchester drives with removable memory disks may also use an interruptive sensor for "disk in place." Some winchester drives utilize an interruptive sensor that is programmed to control the speed of the disk throughout its travel to give uniform information rates (see Figure 17.1).

17.2 FLOPPY DISKS

The track zero sensor is usually an interruptive type sensor that senses the home position. The index sensor is usually a discrete TO-46 IRED and TO-18 phototransistor that senses the rotational speed of the disk. In high density information applications, the movement across the disk may be tracked with a linear optical encoder. Early designs of floppy disks used an optical interrupter sensor to sense the door closure (which locks the disk in place) and allows the motor to operate (see Figure 17.2).

17.3 PRINTER

Printers utilize a varying number of interruptive optical sensors (see Figure 17.3). A daisy wheel type printer utilizes an optical encoder to place the proper character in position for printing. Interruptive type sensors are also used for both right and left margin controls. If the print head has a large mechanical mass, then interruptive type sensors are used to slow the printing speed down prior to reaching the margin control sensors. When a printer ribbon is used a reflective sensor may also be used to identify the "beginning and end" of the ribbon. An interruptive or reflective type sensor is used for verifying the presence of paper to prevent damage to the print head when the paper is not present to cushion

17. Computer Peripheral and Business Equipment Applications   225

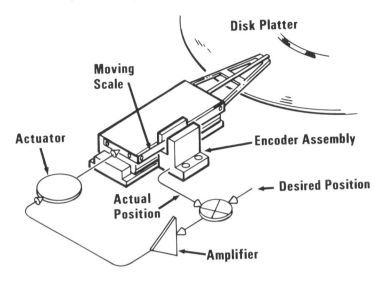

Figure 17.1   Winchester drive track sensing encoder. This application enables the drive head to identify recording tracks. At least three other optical switch applications are possible for Winchester drives.

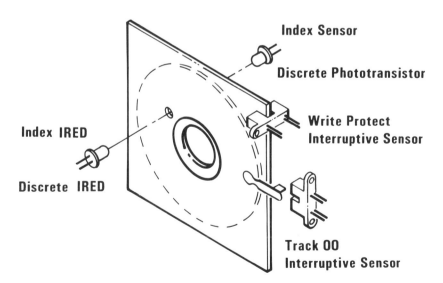

Figure 17.2   Floppy disk optical sensors. The ease of logic interface of optoelectronic switches has made them the technology of choice for computer peripheral mechanical sensing.

Interruptive Sensors

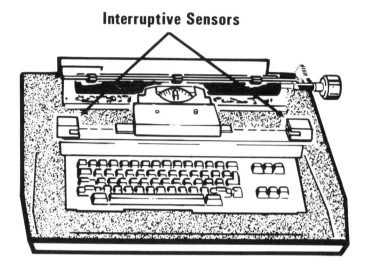

Figure 17.3 Printer optical sensors. Modern electronic typewriter and printer designs are ideal for the use of slotted switches. Printhead control and paper presence are two out of several possible applications worth considering.

the impact. Paper edge sensors are often used to maintain proper paper position in tractor feed applications. Interruptive or reflective sensors may be used with computer paper to track paper movement by viewing the holes on the edge of the paper.

17.4 KEYBOARD AND TOUCHSCREEN

Referring to figure 17.4, as the key is depressed, the energy beam from the IRED to the photosensor is broken. The energy beam may be set up in an X-Y matrix similar to the touch screen in Figure 17.5. The basic operating principle of a touchscreen is to identify the fingertip location by interrupting both "X" and "Y" energy beams. The touchscreen may use discrete lensed IREDs and photosensors, convex magnifying lenses in front of the devices, or timed signals utilizing an IRED focused on a rotating mirror with a fly's eye lens for the receiving photosensor.

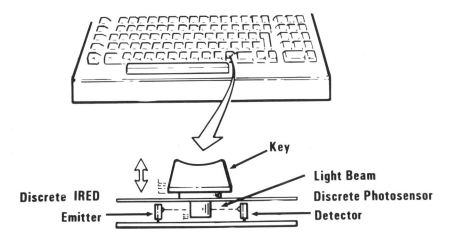

Figure 17.4 Keyboard optical sensors. Optoelectronic keyboards are easy and cost effective to manufacture while offering the advantage of not having mechanical switches to wear out or short.

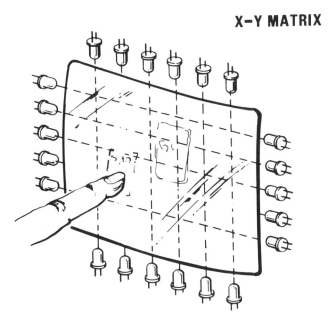

Figure 17.5 Touchscreen optical sensors. The infrared touchscreen has gained popularity among users at a record rate. Optical technology holds an advantage over membrane type touchscreens due to its reliability, accuracy, and ability to leave the screen unobstructed.

17.5 COMPUTER MOUSE

Optical sensors are used in many different ways to input information into computers. Four pairs of interruptive sensors are used in "X" direction and "Y" direction encoders to control direction and amount of movement of pointer on the screen. As the ball mounted in the mouse assembly rotates, the two pairs of optical sensors determine direction and amount of mouse travel and translate this into digital information for the computer. This information may be transmitted over an attached cable that also supplies power, or it may be transmitted in an open air communication link tied to a receiver in the computer (see Figure 17.6).

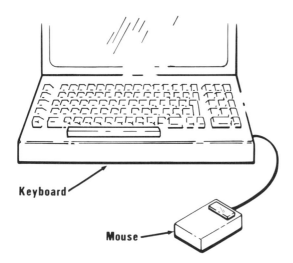

Figure 17.6 A computer mouse for inputting information. The popularity of the computer "mouse" as the nontypist's input tool is at least equal to that of the touchscreen. The internal design is very similar to track ball encoders used in applications ranging from the military to video games.

17. Computer Peripheral and Business Equipment Applications

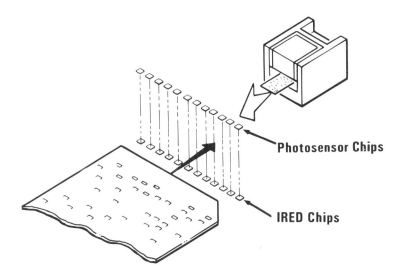

Figure 17.7 80 or 96 column card reader. Perhaps one of the earliest applications of optoelectronic switches was to read computer cards or paper tape.

17.6 CARD READER/TAPE READER

The holes in a computer card are identified by allowing the energy from an IRED chip or discrete packaged device to be received by a mating photosensor. The holes in a paper tape are identified in a similar manner. In addition, the leading or trailing edge of the card can be precisely located utilizing interruptive sensors (see Figure 17.7).

17.7 TAPE DRIVE

Tape drives use a variety of optical sensors. Figure 17.8 shows one technique for controlling the amount of slack. If the tape is being driven from the left and the loop blocks the lower optical pair, then the feed will be slowed or the draw will be slowed. The inverse of this happens when the upper optical beam is opened. In addition to this application, a reflective sensor may be used to detect the reflective mark denoting "beginning" or "end" of tape (BOT, EOT). Interruptive sensors can be used to detect "cartridge in place," "reel size," and "write protect."

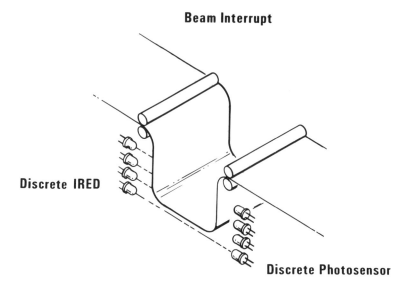

Figure 17.8 Slack length control for tape drive. Computer tape drives offer a variety of areas to apply emitter and sensor pairs. One of the first applications was to control slack so that rapid tape forward and reverse was made possible.

17.8 PHOTOCOPIER

A complex copier may have 30 to 40 infrared optical sensors. Most of these are reflective or interruptive type sensors that track the paper path. They are used to identify the portion of the machine where a jam has occurred. Optical sensors may also be used to control the feed of the paper or the squareness, as well as control the quality of the printing on a real time basis by measuring the density of the toner. An angled transmissive switch can be used to detect "out of paper" as the last sheet leaves the paper bin (see Figure 17.9).

17. Computer Peripheral and Business Equipment Applications

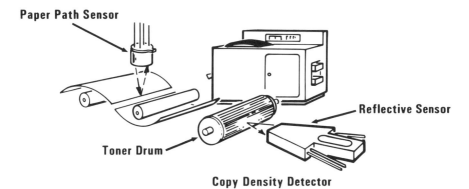

Figure 17.9 Photocopier optical sensing. A large office copier may utilize as many as forty different optical switches. Typically, the entire paper path is monitored.

17.9 MISCELLANEOUS OPTICAL SENSORS

A light pen is nothing more than a VLED or IRED whose energy is focused by external optics or an optical light pipe in conjunction with a photosensor whose viewing area is restricted. It operates as a reflective sensor and discriminates between reflective surfaces such as the light and dark lines in a bar code.

An X-Y plotter may use an encoder setup similar to the computer mouse encoder in principle but with more accurate resolution. The encoder allows both an "X" and "Y" position to be monitored and controlled.

Many of these same type of applications can be carried over to other industries. For example, the banking industry requires optical sensors as paper presence sensors, "door open" sensors, paper squareness sensors, identification and credit card readers supplementing the magnetic readers, and tray location sensors used for the handling of small paper such as checks.

17.10 OPTICAL COUPLERS

Optical couplers (optically coupled isolators) are used to isolate remote terminals from the mainframe computer to prevent ground loops. They are also used in switching power supplies to isolate the control circuits from the transformer primary. For detecting current spikes or shorts that require system shut down, optical couplers are often the first line interface. When the electronic equipment of different manufacturers is tied together to create a network, optical couplers are used to isolate the output or input of the equipment and thus prevent damage due to spurious signals or electrical shorts in part of the network.

18
Industrial Applications

18.1 SAFETY-RELATED OPTICAL SENSORS

Discrete IREDs and photosensors can be utilized to form a safety shield to protect a dangerous area from accidental intrusion (see Figure 18.1). Breaking one or more of the beams would prevent the punch press from operating. This same type of system can be used in similar applications throughout industry.

Handling of explosive material such as gasoline requires the minimization of sparkings. An IRED can be utilized to project an infrared beam down a quartz or plastic rod. This energy bounces across the 45° angles and is received by the photosensor when the external material is air. When the liquid level covers the base of the probe, the energy passes into the liquid, creating a switch action (see Figure 18.2).

18.2 SECURITY AND SURVEILLANCE SYSTEMS

The diagrams shown in Figures 18.3 and 18.4 show a transmissive and reflective system. The transmissive systems may be used up to 500 ft while the reflective systems are limited to one-fourth of the distance of transmissive systems. However, the reflective systems are easier to install since all of the electronics can be located at one point.

18.3 MECHANICAL AIDS, ROBOTICS

One or two sets of discrete opto transmitters and receivers can be utilized to open doors when they are approached and close the doors when an infrared beam is broken beyond them. This same concept is used in Figure 18.6a to measure the height of material breaking the beam. The length of an object can be measured with a similar setup when the object to be measured is moving along a transparent belt. This is shown in Figure 18.6b.

Discrete IREDs and photosensors can be used to limit travel between two points while precise positioning is usually accomplished with a high resolution optical encoder. See Figure 18.7.

PUNCH PRESS
Safety Shield

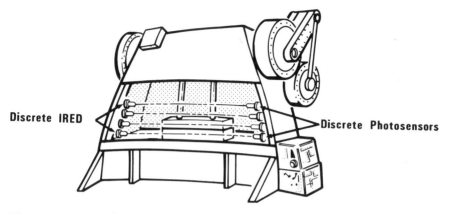

Figure 18.1 Safety shields. This design for punch press operator safety is more reliable than the armstrap method, cannot be defeated easily by the operator, and is more comfortable to use because the operator does not wear the protection.

GAS TANK CONTROL
Liquid Level Monitor

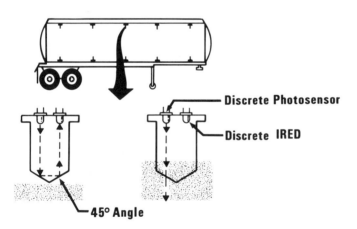

Figure 18.2 Liquid level sensor. This method of level detection is preferred over transmissive types because of its higher reliability. Dirt or other foreign material cannot "clog" the light path to give a false reading, which is particularly dangerous for aircraft fuel safety sensors. Additionally, flammable liquids are safely detected because this sensor type is free of sparks.

18. Industrial Applications

SECURITY SYSTEMS
Reflective Mode

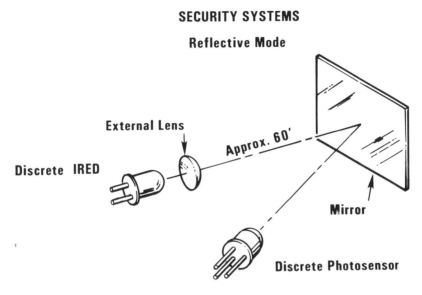

Figure 18.3 Reflective mode optical intrusion sensor. Complete infrared beam type reflective intrusion detectors are manufactured by a variety of suppliers. Installation time for these off-the-shelf units is very minimal.

SURVEILLANCE
Beam Interrupt

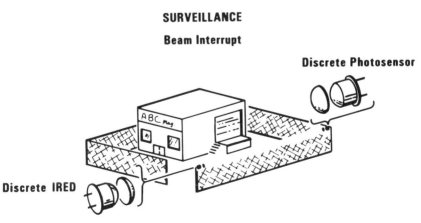

Figure 18.4 Transmissive mode optical intrusion sensor. Perimeter protection using infrared has the advantage of being invisible to the intruder, who never knows he or she is being detected until caught or until an alarm sounds.

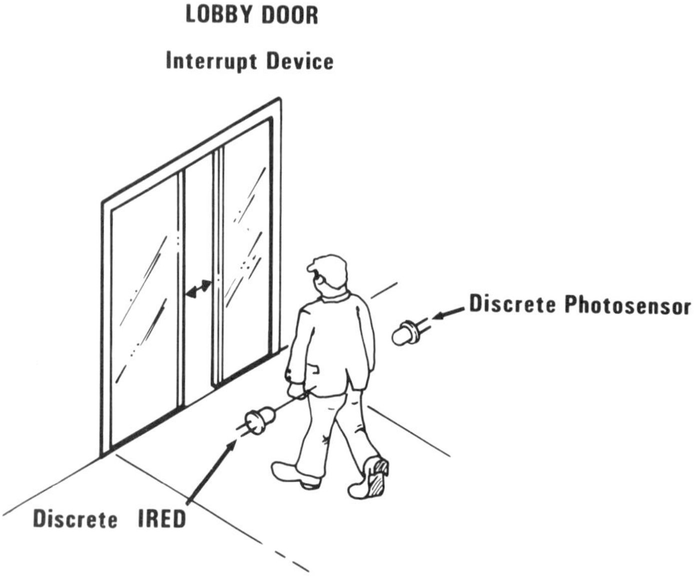

Figure 18.5 Beam interrupt. In areas where the floor mat system is not practical, infrared offers an alternative control method for automatic doors.

18. Industrial Applications

OPTICAL MEASUREMENT SYSTEM

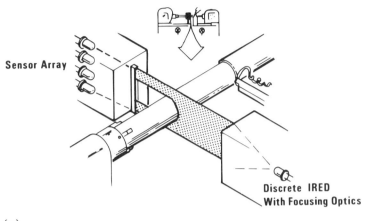

(a)

PIECE PART MEASUREMENT

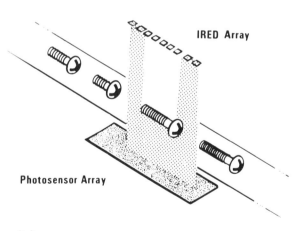

(b)

Figure 18.6 (a) Method of measuring machined diameter. Measurement within discrete amounts, such as for a process control, is possible using carefully aligned arrays of IREDs and sensors. (b) Method of checking an object's length. In this example, one quality measure, length, is checked using an optoelectronic array.

ROBOTS

Shaft Position Encoder
and EOT/BOT Sensors

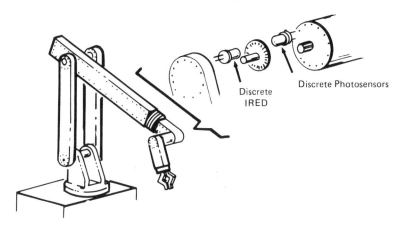

Figure 18.7 End of travel, beginning of travel, and precise location control encoders. The growth of robotics has partially fathered the growth of optical linear and rotary position encoders.

DIGITAL CALIPERS

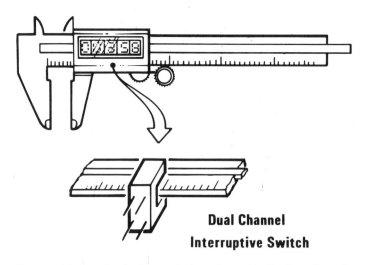

Figure 18.8 Dual channel encoder. Both the direction of movement and the position of these electronic calipers is recorded. Resolution to 0.001 in. is possible.

18. Industrial Applications

PDIP HANDLER

Beam Interrupt

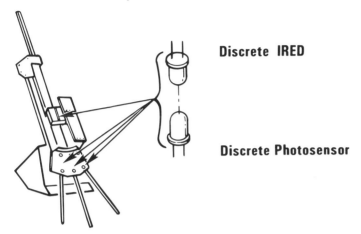

Figure 18.9 Automatic aids for plastic dual-in-line package IC handling. One of the first industries to quickly incorporate infrared optoelectronic position sensing and counting into its production equipment was the electronics industry.

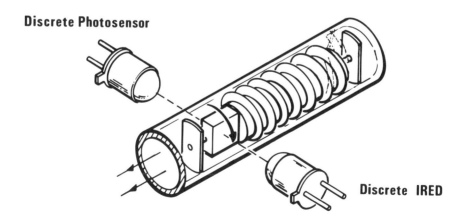

Figure 18.10 Measurement of fuel flow. There are no seals to fail or sparks to reach explosive fluids using this design to measure flow.

A dual channel interruptive switch can be used to determine both direction of travel and relative position. Normal design resolution is 0.003 in., but certain applications have been able to improve this to 0.001 in. See Figure 18.8.

An interruptive optical sensor fabricated from discrete components can be used to determine when a transparent carrier of PDIPS needs to be replaced. See Figure 18.9.

Fuel or liquid flow can be measured by counting the revolutions per unit of time in a known volumetric area. See Figure 18.10.

18.4 MISCELLANEOUS OPTICAL SENSORS

Interruptive type sensors can be used to detect thread breakages in textile mills. A reflective sensor can be used to turn on a sewing machine when the cloth to be sewn approaches the needle. An encoder can be used for programming machine tools and for remote reading of gas pumps.

Photosensors have been used to speed up production in beer factories. An infrared emitter is placed inside a beer can as an external array of sensors looks for holes or leaks. This is a much faster system than the previously used pressurization of the cans.

Industrial smoke detectors are similar to the ones used in the home. An IRED is pulsed and projects its energy into a chamber. When smoke particles are present the energy is reflected to a photosensor, which trips an alarm when the level is exceeded. The industrial unit may actually measure the amount of smoke.

18.5 OPTICAL COUPLERS

The zero current and voltage switching versions of the optically coupled isolator with triac driver output find wide application in industry. DC controllers that have to actuate an AC line are numerous (motor speed controllers, traffic signal light, and solid state relays). See Figure 18.11.

18.6 CONTOUR SENSORS

Critical contours can be checked automatically following fabrication by using an array of reflective sensors. If the product (e.g., body part panel) does not meet specifications, then it will not lie flat against the test fixture. The energy reflected off the misaligned object reaches the photosensor and signals the failure to match the contour of the test fixture. See Figure 18.12.

PROGRAMMABLE CONTROLLER
Optically Coupled Triac Driver

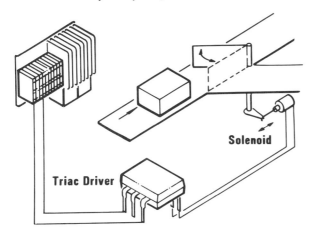

Figure 18.11 Optically coupled isolator with triac driver output. As a protective device, the controller or micro-processor is isolated from the higher voltage and current used to power the solenoid.

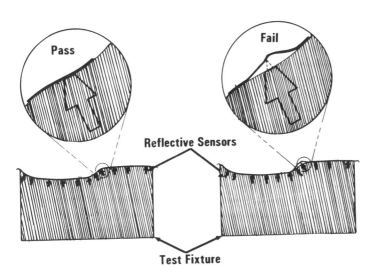

Figure 18.12 Contour checking of a metal stamping. In this application, multiple reflective sensors carefully mounted in the test fixture can check hundreds of contour points instantly, assuring a perfect end product every time.

19

Automotive Applications

19.1 EXISTING APPLICATIONS

When a load is added to the car equipped with the automatic load leveler, the flag that normally will break the upper beam but not the lower beam drops, causing the lower beam to be disrupted. Air is pumped into the appropriate shock absorber until the lower channel is clear. When the load is removed, the flag raises and opens the upper energy beam. Air is released from the shock absorber until equilibrium is again reached. It should be noted that due to the high optical contamination in this area, this system is being phased out and replaced with magnetic sensors. The operating principle remains the same, but dirt and grime do not affect the reliability of the magnetic system. See Figure 19.1.

A flag with a reflective surface on either end is mounted to the speedometer cable. As the flag rotates, the sensor will output two signals per revolution. By summing these signals for a fixed period of time and comparing them with a preset number, one can control the speed of the car. This is a simple tachometer type of application. The automotive industry is moving towards elimination of the speedometer cable, instead of receiving speed information directly from the drive shaft. Therefore, this sensor function will be replaced in future automative applications wth a different type of sensor. See Figure 19.2.

19.2 FUTURE APPLICATIONS

The ABC sensor discussed in Part II can be used to automatically adjust the brightness of the panel lights to a preset contrast level. As the automobile moves from city to country, ambient light levels drop, and the displays are adjusted to a preset readability level. The viewer is usually not conscious of the illumination change. See Figure 19.3.

There are many other potential applications for optical sensors in future automobiles. They are generally associated with passenger compartment luxury options that take advantage of the less hostile environment.

19. Automotive Applications

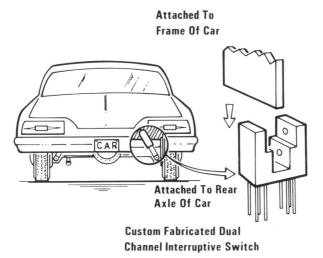

Figure 19.1 Automotive automatic load leveler. This application is effective in principle and was actually produced for many years. However, newer systems that are unaffected by dirt and grime are replacing this style of load leveler.

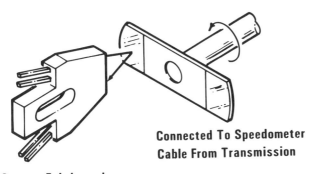

Figure 19.2 Automatic cruise control. The heart of many cruise control systems is the optically based speed sensor. This may also be used to feed information to an electronic speedometer, if desired.

AUTOMOTIVE

Dashboard Illumination Control

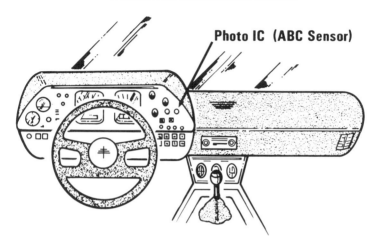

Figure 19.3   Panel lamp brightness control. Contrast is held constant by the photo IC used to control the dashboard illumination.

1. Door ajar (interruptive or reflective sensor)
2. Trunk lid ajar (interruptive or reflective sensor)
3. Hood ajar (interruptive or reflective sensor)
4. Oncoming headlight sensor (photo IC)
5. External lights operative (photosensor with possible fiber cable linkage)
6. Oil contamination (interruptive sensor with pulsed IRED and fiber optics to pick up point)
7. Short range remote communication link:
   a. Releasing door locks by coded signal
   b. Communication within passenger compartment: door locks, window raise or lower, seat adjustment, mirror adjustment, other remote switch control (rear window defoggers), hood opener, trunk opener, radio speaker sound transfer
   c. Data link within steering column for steering wheel mounted displays and controls
8. Long range remote communication link:
   a. Car starting
   b. Air conditioning or heater on/off
   c. Windows up or down
   d. Lights on (locating of car or safety for intruders)

19. Automotive Applications

    e. IRED (radar viewing) used to locate vehicles in viewing dead zone
9. Touchscreen panel switch control for dash
10. Sensor for determining driver drowsiness
11. Sensor to determine presence of glare ice on road
12. Sensor for determining buildup of ice in wheel well on interstate to prevent limited turn radius when existing

The above list is by no means all inclusive. For each possible application, cost-benefit trade-offs will determine the actual feasibility. For example, while cargo bay door sensors are considered mandatory equipment for all commercial airliners, automotive open trunk indicators will not be considered practical for all vehicles until the cost reaches a very nominal amount.

20

Military Applications

20.1 MILITARY APPLICATIONS FOR OPTICAL SENSORS

Optical sensing applications in the military environment encompass many of the same functions that exist in the commercial market. Since most of the components used have more stringent processing, specifications such as hermetic closure and extra environmental testing lead to significantly higher costs. Nevertheless, in the life and death situations of the military, reliability supersedes cost. For example, a simple interruptive switch can prevent a homing torpedo from seeking out the firing vessel. If the torpedo were to turn 180° and head for the firing vessel, a flag would interrupt the energy beam and prevent the torpedo from continuing on that course.

Shaped charges are placed around a hypersonic missile. Flight patterns can be changed by selective explosion of these charges. Optical couplers are used to isolate these firing circuits from the control electronics. See Figure 20.1.

Optical couplers are also used in switching power supplies in both ground control and flight hardware. They are used to isolate the control circuitry from the primary or the detect current spikes or shorts that would require system shut down.

The need for panel light brightness control in aircraft is far greater than in automobiles both due to the number of instruments and a more rapidly changing environment. The ABC sensor, with its capability to multiplex, offers an excellent cotroller for this type of application. See Figure 20.2.

Other unique applications in the military environment include the following:

1. IRED illumination for helicopter landing or running lights. A TO-46 IRED operated at 100 mA and used with infrared conversion goggles will generate adequate energy to read a newspaper in a conference room with no other lighting. Such devices are also critical for black-out formation flying when used as running lights.

2. PIN quadrant diodes are used as homing sensors for guidance control. The photosensitive area is divided into four quadrants and, by use of a finely tuned infrared spot size, can perform automatic guidance. The control system forces the spot to the center or zero signal area.

20. Military Applications 247

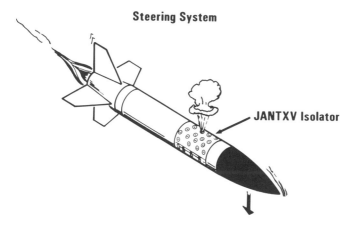

Figure 20.1 Steering system of a hypersonic missile. Catastrophic failure of control circuitry is prevented by optically isolating the higher voltage firing circuits.

AIRCRAFT

Automatic Panel Brightness Control

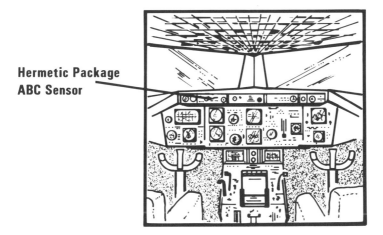

Figure 20.2 Automatic brightness control sensor. In this application, brightness could still be adjusted by the pilot; however, once set, contrast would be maintained regardless of changes in ambient light.

3. Fighter plane pilots in high gravity environments can utilize a reflective sensor mounted adjacent to the eye for supplemental aircraft control. By moving the eye, the pilot can generate coded signals, which can then be used to control certain aircraft functions.

4. A visible LED array mounted on a vibrating member can generate signals similar to a miniature TV screen on a localized area of eyeglasses. This "head up display" information could be used to keep the pilot informed of basic gauge position while his eyes are not watching the instrument panel.

5. Digital altitude information can be obtained from an encoder mounted within the normal altimeter. This information could be read from a ground station without direct communication to the plane operator.

21
Consumer Applications

21.1 TV GAMES OR TOY CONTROLS

Remote control of toys or TV games with optical sensors is quite common. The signal pulses can either be transmitted through a connecting cable or from a pulsed IRED.

The basic operating principle of a joystick as shown in Figure 21.1 is quite simple. Four interruptive type sensors are located at the four compass points. As the joystick is moved, one of two of these infrared beams are broken. The movement of the controlled dot on the screen is then initiated in accordance with the signals from the sensors. If the "north" energy beam is broken, the dot will move north at a fixed rate. If north and east are broken, the dot will move northeast. The direction is controlled but the rate of movement is fixed.

The trackball (see Figure 21.2) operates similarly to a "mouse" except the ball is rotated by the hand. Two dual channel interruptive sensors are located for "X" and "Y" movement. The dual channel allows quadrature to be generated, which controls forward or reverse. Counting the pulses per unit of time elapsed from one channel in "X" or "Y" can generate velocity. The advantage of the trackball over the joystick is the addition of velocity control.

21.2 TV AND CATV REMOTE COMMUNICATIONS

Ultrasonics was originally used in hand-held transmitters for remote control of TV sets (see Figure 21.3). This has been replaced by pulsed IR energy, which allows more complex information to be transmitted.

The coded information can be sent by transmission of frequency "1" for a logic "1" or frequency "2" for a logic "0." A second type of coding uses pulse frequency as the code. This same system could also be used by a waiter communicating the coded order to the bar and kitchen for improved service. Currently, it is used for remote control of TV cameras, some garage door openers, remote meter reading, and even swimming pool and hot tub control.

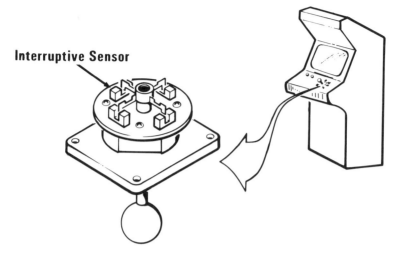

Figure 21.1 Joystick optical sensor. The ultimate test of a switch lies in the video game parlor. Nonwearing slotted optical switches are ideal for this application because of their direct logic compatibility and reliability.

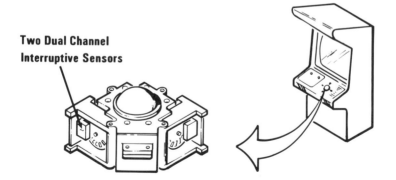

Figure 21.2 Trackball optical sensor. Similar in concept to the computer "mouse" but more durable in design, the trackball encoder also has military and industrial control applications.

Wireless Communication

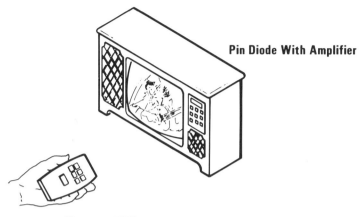

Figure 21.3   Remote control of TV and CATV systems. By far the most popular application of open air infrared communication, television remote control technology is discussed in detail in Chapters 13 and 14.

RECORD OR VIDEO DISK PLAYER

Playback System

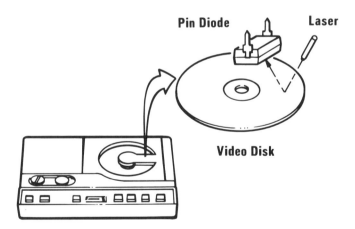

Figure 21.4   Optical track control. In addition to basic playback, several additional functions are provided by this optical feedback system. For conventional players, optical sensors may detect disk size, play completion, and other events.

251

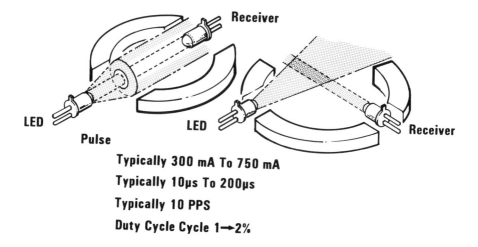

Figure 21.5 Optical smoke detectors. Perhaps one of the highest volume users of discrete infrared emitters and sensors, smoke detectors have saved lives all over the world. Their principle of operation is one of the simplest optoelectronic applications.

21.3 VIDEO DISK OR RECORD CONTROL

The laser beam is reflected from the grooves in the record or video disk. The change in reflected signal allows the tracks to be identified. On a record player, an interruptive type sensor can be used to detect when the tone arm has completed its arc and thus the record is complete. In a video disk, interruptive sensors can be used to determine when the disk has been inserted (see Figure 21.4).

21.4 DOLLAR BILL CHANGER

Many money changers operate on an optical sensor principle. Calibrated IRED/photosensor pairs are located at key points on the bill. A good bill is detected by the transmission through the bill at these key points.

21.5 SMOKE DETECTORS

Most optical smoke detectors operate on the principle of detection of reflected energy from the smoke particles. They can be used, however, where the presence of smoke will block the IR energy (see Figure 21.5).

21. Consumer Applications   253

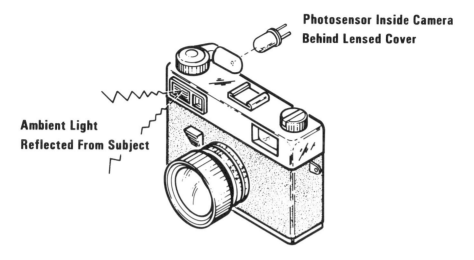

Figure 21.6 Camera use of optical sensors. In addition to simple exposure control, infrared focusing is also possible but requires precise and complex optical design in order to be functional.

21.6 COIN CHANGERS AND SLOT MACHINES

A coin changer offers a number of different applications for interruptive type optical sensors. If the coin changer must identify nickels, dimes, and quarters, as many as thirteen sensors may be used. In order to control the availability of coins for making change, a coin "empty" and a coin "full" sensor is required for each denomination. When the change slot is full, a coin diversion sensor is required to divert the coin to the reservoir box. Three more sensors can be used to read the elapsed time from leading edge to trailing edge and thus control size. The thirteenth sensor is used to establish that the coin has dropped and stayed in the machine. Many of these same applications are used on slot machines.

21.7 CAMERA APPLICATIONS

A camera can use the light from the outside to control the shutter open time and/or the time the flash is "on." A diode photosensor will integrate the amount of light received and at a predetermined point can close the shutter or cut off the flash (see Figure 21.6).

IREDs are sometimes used to expose IR sensitive film. This is useful when pictures are desired, but the flash can be disturbing to adjacent people (such as in a slide show or theater).

21.8 OPTICAL GOLF GAME

Interruptive type sensor can be used to measure velocity and direction of a golf ball. Slides can be taken from various points on a golf course. The first slide depicts the view from the tee on hole one. The golfer drives the ball and optical sensors measure the velocity and direction. The ball then strikes the screen, while the computer calculates where the ball would come to rest. The appropriate picture then comes up. The ball is replaced on the tee and the golfer swings again. The cycle repeats until all 18 holes are completed, rain or shine.

21.9 POOL CUE BALL SENSOR

Coin operated pool games operate with a larger cue ball than object ball. When the cue ball goes into a pocket, the larger size is detected and the cue ball is brought back to a separate outlet. Since realism is lost by this system, an optical system could be used that would detect a different cue ball reflectivity. The cue ball surface could be more or less reflective than the object balls.

21.10 HOUSEHOLD APPLIANCE CYCLE CONTROL

Many household appliances are switching to solid state timers and controls. This opens up many new applications for optical couplers that can operate from a DC input and be used to control resistive, capacitive, or inductive AC loads (see Figure 21.7). Solid state switching for home appliances is a very rapidly growing technology.

Figure 21.8 Eye shutter synchronization using infrared for three-dimensional television. Scanning 60 frames per second, including 30 left eye images and 30 right eye images, would make three-dimensional video a possibility. The electronic eyeglasses, capable of shuttering at 30 times per second and synchronized by the IR beam, have yet to be perfected. On the other hand, generation of the required video images is a relatively easy process.

21. Consumer Applications   255

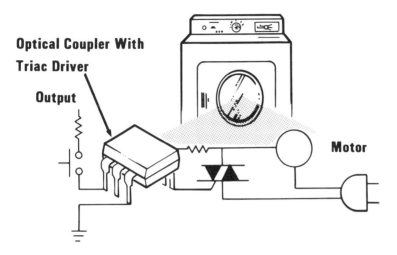

Figure 21.7 Optical coupler used to control AC loads. The increasing concern for consumer safety along with the increased use of microprocessor controls has led to the incorporation of optically coupled isolators in many homes appliances, including mixers or blenders, dishwashers, and clothes washers and dryers.

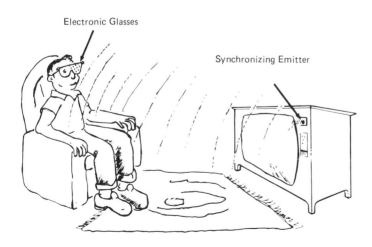

21.11 THREE-DIMENSIONAL VIDEO OR FILM

Three dimensional imaging, long a dream of engineers, but with little real practical success, may soon be possible for television or film. The secret is getting the left and right eye images separated. Color filter or polarizing setups would no longer be needed if special glasses with electronic light shutters could be made practical for widespread use. The glasses would be synchronized with alternating left and right eye television images by use of an infrared carrier beam received by a PIN diode located in the frame of the glasses (see Figure 21.8).

22
Medical Applications

22.1 INTRODUCTION

The mention of optosensors with respect to medical electronics brings to mind the counting of pills, the verification of cleanliness, or that the pill bottles themselves are full. Additional applications include the protection of the patient from electronic monitoring equipment through the use of optocouplers, or the protection of the doctor by optocouplers when the patient's hear is shocked and started. There are many more sophisticated applications that are utilized in medical optoelectronics.

22.2 INTRAVENOUS DROP MONITOR

Intravenous injection rate can be easily be controlled by an interruptive sensor. The curved leading edge of the drop bends the IR energy away from the sensor. By summing the output pulses for a given time interval control of rate can be obtained (see Figure 22.1).

22.3 PULSE RATE DETECTION

A reflective sensor located adjacent to a blood vessel can detect pulse rate. The change in the blood vessel when the heart pumps causes a change in signal level in the reflective sensor. This system is used in fingertip pulse rate monitors for joggers and other hand held monitors. Analysis of blood samples is also aided by interruptive type sensors.

22.4 MEASUREMENT OF GLAUCOMA

Testing for glaucoma is nothing more than measuring the pressure of the fluid in the eye (see Figure 22.2). A metered air jet is directed into the eye. The pressure is controlled by an interruptive type sensor and a moving ball across the aperture. When this air

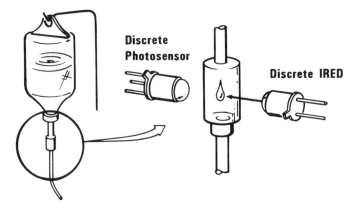

Figure 22.1 Optical interruptive sensor for controlling flow rate. Infrared intravenous drop monitors are now standard hospital equipment. Not only is the medication flow monitored, but the equipment allows quick and accurate adjustment to the desired flow rate.

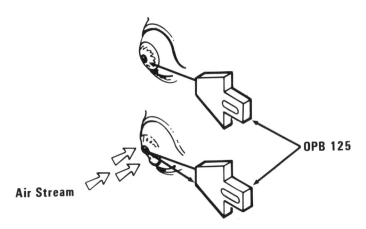

Figure 22.2 Testing for glaucoma using optical sensors. Glaucoma patients are at risk of blindness due to excessive internal fluid pressure within the eye. Early detection, the key to a successful cure, is made painlessly possible with this infrared detection method. The test has become a routine part of normal eye examinations.

22. Medical Applications

pressure reaches the point that the front of the eye is flattened, an adjacent IRED energy beam bounces off the flattened eye and is detected by an adjacent photosensor. The air pressure of the jet is recorded at this point, and the corresponding internal eye pressure is then detemined.

22.5 OPTICAL SWITCH FOR PARAPLEGICS

A reflective sensor is mounted adjacent to the eye. The signal is obtained through the controlled movement of the eye. This signal could be coded by the number of movements or by stopping a sequence of lights that control various functions. For example, these simple functions might be:

1. Call nurse
2. Raise or lower bed
3. Change TV channels

This would allow a patient with only eye movement to control simple functions.

23

Telecommunications

Optical couplers have had wide use in telecommunications in preventing ground loops where common equipment would be located in a variety of locations. In the future, optical sensors will play a large role as service policies change and convenience options and equipment functions are expanded (see Figure 23.1).

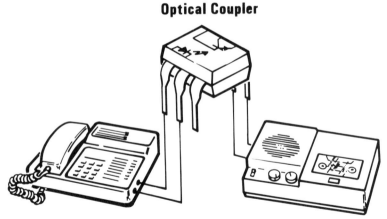

Figure 23.1 Ground loop isolation. One of the most popular uses of optoisolators in telephone and telephone accessory equipment is the ring detector to sense, yet block the high voltage ring signal. Other uses include network interconnection for equipment electrical protection and elimination of ground loops.

23. Telecommunications

Interruptive switches will replace the mechanical switches identifying when the handset is returned to the base station. Short distance remote communication can be used to eliminate the telephone cord between the handset and the base station. Optical coupler use will increase as the basic communication system converts from linear to digital transmission.

Glossary

Acceptable quality level (AQL): The maximum allowable average percentage of defective components that a supplier is permitted to present for acceptance. Also referred to as *acceptance quality level*

Acceptance angle: The angular limit off of the optical axis for which a photosensor will still detect an energy ray to give at least a half power level signal.

Acceptance cone: A cone defined by the acceptance angle such that a ray within the cone will be above the half power level.

Alignment: The process of positioning emitter and sensor for maximum infrared transmission (optical alignment) while placing the beam in the desired location. Infrared sensitive scopes or luminescent materials are often used as aids for aligning complex infrared optical systems.

Ambient temperature: The temperature of the gas or liquid surrounding a component.

Alloy: A composition of two or more materials, usually metals, combined for the purpose of improved performance characteristics.

Angle of emission: The angle at which a ray exists an optical package, measured from the optical axis.

Angstrom: A unit of length equal to 10^{-10} m, sometimes used to quantify the wavelength of electromagnetic radiation.

Anode: The positive side of a diode to which a positive voltage must be applied to facilitate a forward current.

Aperture: An opening used to control or limit the transmission of electromagnetic radiation.

Aperture angle: Used interchangeably with beam angle. It is usually better to specify whether a given measurement refers to the angle measured from the optical axis to the half power point (half angle) or between the two half power points (included angle).

Axis of measurement: The direction from which ratiometric measurements are taken.

Alpha: Used in electrical engineering to refer to emitter-collector gain (transistor connected as a common base amplifier). In a junction transistor, alpha is less than one.

Base: The P portion of an NPN transistor, and in the case of a phototransistor, the photosensitive region. For a PNP transistor, the N region is the control terminal or base.

Beam angle: Used interchangeably with aperture angle. It is usually specified as the included angle from one half power point to the other.

Beam half angle: The angular displacement of emitted electromagnetic radiation from the optical axis to the point of half maximum power.

Cathode: The negative region of a PN diode to which a negative voltage or ground must be applied for forward current.

Collector: The terminal of a transistor to which a bias voltage (positive for NPN types) is normally applied.

Collector-base breakdown voltage: The voltage at which current suddenly rises from normal leakage to a higher, specified level. Positive voltage would be applied to the collector, the emitter would be open, and the base grounded or negative for this condition.

Collector current: A measure of the current flowing through the collector, expressed in amperes.

Collector-emitter breakdown voltage: The voltage at which current suddenly rises from normal levels to a higher specified level. The transistor is normally biased with no base input.

Collector-emitter saturation voltage: The voltage measured across the collector to emitter with a specified optical or electrical base input and output load.

Color temperature: The physical temperature of a blackbody emitting radiation peaked at a wavelength corresponding to that emitted by a nonblackbody radiator.

Common emitter: The grounded emitter configuration where the emitter terminal is common to both input and output.

Critical angle: The angle of incidence at which light is no longer transmitted through an n1/n2 interface but is instead reflected by the interface.

Current sensing: The application of an optoisolator to detect current (through the IRED) in a circuit, such as a power supply or telephone network.

Current transfer ratio: Output current divided by IRED current under specified test conditions (a ratio).

Dark current: A measure of the current in a phototransistor or other photosensor with no radiation present.

DC current gain: Collector current divided by base current [H_{FE} (DC), h_{fe} (AC); a ratio].

Glossary

DC isolation current: The current between input and output at a specified test voltage with all input leads shorted together and all output leads shorted together.

DC isolation voltage: The rated (and tested) input-to-output DC isolation voltage limit with input leads shorted together and output leads shorted together.

Diode: A two-terminal semiconductor device that conducts freely in one direction only.

Duty cycle: Pulse width divided by total cycle time, usually expressed as a percentage of time that the device is on.

Duty factor: Duty cycle expressed as a ratio.

Efficiency: Referring to a photodiode in photovoltaic mode, the ratio of maximum power output to total incident radiation energy.

Emission angle: See Beam angle.

Emitter: (1) The negative voltage terminal of an NPN transistor under normal operating bias. (2) Used to refer to an infrared emitting diode. (3) Refers to any source of radiation (i.e., infrared emitter).

Emitter-collector breakdown voltage: The voltage at which emitter-collector current rises to a nondestructive specified level that is significantly higher than normal leakage current.

Emitter current: (1) A measure of the current flowing through a transistor's emitter terminal. (2) The current through an IRED.

Fall time: A measure of the time required, under specified test conditions, for a waveform to drop from 90% to 10% of its original level.

Focal plane: The plane at which rays converge to form a focused image.

Interrupter: Synonym for transmissive assembly.

Interrupter module: Synonym for transmissive assembly.

Operating life: The minimum length of time that a device may safely be expected to perform within established specifications.

Optical axis: The line designated as such and used as a measurement reference; usually the locus of emitted or received rays, perpendicular to the focal plane.

Optical coupling: Energy transfer from a source to a photosensor.

Optical matching: Refers to similar optical package designs for IREDs and photosensors.

Optocoupler: A device designed to provide electrical isolation through conversion to an optically coupled signal and back to an electrical signal.

Optoelectronics: A term used for electronic systems or components that interface with optical systems or components.

Optoisolator: Synonym for optocoupler.

Parabolic reflector: A reflective disk-shaped sheet formed such that a cross section corresponds to a mathematically calculated parabola. A parabolic reflector will have a clearly defined central focal point useful for accurately collimating a point source or focusing collimated light to a single point.

Peak wavelength: The wavelength for which the emitted power is greatest; i.e., the wavelength of maximum power output.

Phase: A concept used to measure the relative positions of separate waveforms.

Photoconductive: A term referring to increased conductivity as a function of increasing incident radiation.

Photocoupler: Synonym for optocoupler.

Photocurrent: Current that is generated in a photosensor as a result of incident radiation.

Photodarlington: A photosensor consisting of a Darlington transistor (i.e., two cascaded transistors together on a single piece of semiconductor material) with a photosensitive base region in the first transistor.

Photodetector: A general term usually referring to a photodiode, phototransistor, photodarlington, photo IC, or other device that responds electrically to radiation.

Photodiode: A two-terminal semiconductor device designed to conduct when exposed to incident radiation.

Photoemissive device: Synonym for emitter or IRED.

Photoemitter: Synonym for emitter or IRED.

Photon: A single unit of electromagnetic radiation; equal in energy to the product of frequency and Planck's constant.

Photosensitive device: Synonynm for photosensor.

Photosensor: A general term usually referring to a photodiode, phototransistor, photodarlington, photo IC, or other device that responds electrically to radiation.

Phototransistor: A transistor designed with a photosensitive base region so that conduction occurs when radiation is incident upon the base.

Phototriac: An integrated circuit designed to pass current in both directions when incident radiation exceeds a specified threshold.

Point source: A single point that is the starting place for all emitted rays; however, in actual practice optoelectronic design may treat an emitting area as a point source as long as specified ratios are not exceeded.

Polar coordinates: A system for location using distance and angle as point coordinates which are sometimes used to plot emission patterns.

Power: A measure of the rate at which work is done, generally expressed in watts.

Power derating: The system used to correlate the cooling effect possible at a given ambient temperature with the power dissipation capability of a device.

Radiance: The intensity of the energy passing through an area, divided by the measure of the area.

Glossary

Radiant flux: The measure of radiant power, usually expressed in watts.

Radiant intensity: The measure of radiant flux per unit of solid angle, usually expressed in milliwatts per steradian.

Radiant incidence: The measure of radiant flux incident upon a surface; i.e., milliwatts per square centimeter.

Radiation pattern: A description of radiant incidence of radiant intensity as a function of position from the source, usually in a single plane.

Ray path: An imaginary line perpendicular to a wavefront which depicts the path of the wave.

Reflective assembly: Synonym for reflective sensor.

Reflective sensor: An assembly consisting of IRED and photosensor used to detect objects by reflecting an infrared or light beam off the object.

Responsivity: A measure of the sensitivity of a photodetector equal to output current divided by radiant flux incident upon the sensing surface.

Reverse bias: For a diode junction, interconnection in the nonconductive mode.

Reverse current: The current that flows in the reverse bias condition.

Rise time: A measure of the time required, under specified test conditions, for a waveform to increase from 10% to 90% of its peak level.

Rotary encoder: A position sensor that signals rotary motion. Optical rotary encoders may detect speed, direction, and absolute position, depending on complexity.

Silicon: The base material for photosensors and other semiconductor materials when used in relatively pure crystalline form with specific materials added.

Slotted switch: Synonym for transmissive assembly.

Snell's law: The principle that predicts the change in ray path as radiation crosses an optical interface; namely, $n_1 \sin \theta_1 = n_2 \sin \theta_2$, where n_1 and n_2 are respective indices of refraction and incident angles are measured from a normal to the interface.

Spectrum: A term used to refer to the broad range of wavelengths within a given category of radiation; i.e., visible spectrum.

Spectral bandwith: The range of wavelengths between outer half power points for a given source.

Spectral bandwidth: The range of wavelengths between outer half wavelengths.

Steradian: A solid angle that intersects a surface area equal to the squared radius of a sphere.

Storage temperature: The maximum and minimum safe temperatures for storage of a device without damage occurring.

Supply voltage: The voltage required to operate a circuit.

Transistor: Semiconductor device with two junctions and three terminals known as collector, base, and emitter.

Transmissive assembly: An assembly utilizing an infrared ray path from IRED to photosensor to detect the presence of an object that breaks the infrared ray path.

Transfer ratio: See Current transfer ratio.

Trigger level: As applied to photosensors, the levels of radiant intensity at which the device turns on or off.

Trigger voltage: The voltage levels that trigger a device to turn on or off.

Visible emitter: Synonym for VLED.

Visible LED: Synonym for VLED.

VLED: A diode that emits visible radiation when forward current passes through its junction.

Wavelength: A measure of the length of an electromagnetic wave, calculated by dividing the speed of light by the specific frequency of the light, but usually measured empirically.

Bibliography

J. Brophy, *Basic Electronics for Scientists (2nd)*, McGraw-Hill Book Company, New York, 1972.

A. Chappell (editor), *Optoelectronics Theory and Practice*, Texas Instruments, Ltd., Bedford, England, 1976.

The Electronic Engineer, A Course on Optoelectronics, Chilton Company, Bala Cynwyd, Pennsylvania, 1971.

S. Gage, D. Evans, M. Hodapp, H. Sorensen, *Optoelectronics Applications Manual*, McGraw-Hill Book Company, New York, 1977.

S. Gage, D. Evans, M. Hodapp, H. Sorensen, D. Jamison, B. Krause, *Optoelectronics/Fiber-Optics Applications Manual (2nd)*, McGraw-Hill Book Company, New York, 1981.

E. Hecht, A. Zajac, *Optics*, Addison-Wesley, Reading, Massachusetts, 1974.

JEDEC Standard No. 77, JEDEC Solid State Products Engineering Council, 1981.

J. Wilson, J. Hawkes, *Optoelectronics: An Introduction*. Prentice/Hall International, Inc., Englewood Cliffs, New Jersey, 1983.

The following application bulletins, published by TRW, Inc., were used with permission in the preparation of this text:

A.B. 105, Thermal Behavior of GaAs LEDS, Cognard, W. Nunley.
A.B. 108, Motion Sensing with Optical Interrupters, T. Sward, and W. Nunley.
A.B. 111, Soldering to Semiconductor Leads, W. Nunley and H. Brown.
A.B. 112, Two Channel Optical Interrupters, V. Dahlberg and W. Nunley.
A.B. 113, Reflective Assemblies, T. Sward
A.B. 114, Gallium Aluminum Arsenide, D. Wolfe.
A.B. 116, Linear and Rotary Encoders, J. Davidson and L. Johnson.
A.B. 118, Understanding Infrared Diode Power Ratings, K. Bailey.

A.B. 119, A Comparison of Plastic versus Metal Packaging for Infrared Sensors and Emitters, M. McCrorey.
A.B. 120, Designing a Wide Gap Optical Switch, T. Eichenberger.
A.B. 121, Understanding the Dissipation Rating of the Optical Semiconductor, M. McCrorey.

Index

ABC,114-121
Aircraft fuel level detector, 157-159
Aircraft panel brightness control,244,247
Aperture,149-150
  etched,150
  encoder,139
  molded,149
  numerical,194
  power measurement,62
  width,135
Apertured radiant incidence, 59
Appliance control,254
Automatic brightness control,114-121
Automatic gain control,146
Automatic shutter control, 253
Automation,20-21

Bar code,231
Base width,84
Beam angle,32
Beam pattern,33-34,36,57
Beer can,240
Beta (see Current gain)
Bill changer,252
Bipolar technology,99-109
Bonding:
  thermocompression,20-22
  thermosonic,21
  ultrasonic,21
Buried layer,103
Burn-in,63

Card reader,229
Carriers,4,69
  acceptor,7
  donor,7
  electrons,69
  holes,69
  N-type,4,7,101
  P-type,4,7,101
Catastrophic failure,65
Change counter,252
Characterization,26
  coupled,60
  forward current,26
  phototransistor,82
  storage temperature,37
  temperature,28
Chip:
  mounting,19
  placement,33,35
  sizes,13,18
Coating,93
Coin detector,252,253
Collector current:
  vs. irradiance,93
Computer mouse,228
Contact design,28
Contour checker,240-241
Cordless telephone,261
Coupled characteristics,60
Coupler (see Optical isolator)
Cover art explanation,128
Critical angle,18
  after contouring,16
  description,16
  fiber optic,193-196
  in housing,149

IRED,16
  liquid level sensing,159, 233-234
  paper sensing,17
  after roughening,17
Cross talk,162
Cruise control,242-243
Current gain,84
Current limiting resistor, 213
Current surge detector,246
Current:
  gain,90
  vs. irradiance,93
  leakage,87
  photo,69,84
  sensor output,59
  transfer ratio,188

Daisy wheel,224
Dark current,75
  leakage,80
  phototransistor,87
  switching,147
  vs. temperature,79,91
Darlington:
  photodarlington,222
  transistor,84
Dash board display,244
Degradation,35,62-65
  compensating for,215
  in couplers,172
  dark line defects,53,62-65
  in encoders,146
  pulse mode,202
  vs. temperature,63
  vs. time plots,55
Derating curve,46,51-52
  calculations,52,202-210, 215
Digital altimeter,248
Digital calipers,238
Digital potentiometer,249-250
Diverging angle,205-206
Dollar bill changer,252
Door sensor,224,231,233, 244
Doping:
  aluminum,10-11
  concentrations,10
Duty cycle,202,204

EOT/BOT sensor,229
Efficiency:
  conversion,5
  coupling,5
  gap,5
  optical,14
  package,18
  photon,53
  quantum,75-77,93
  side radiation,18,58
Electron volts,5
Elevator door sensor,233,236
Emitters, 3-13
  circuitry,213
  dimensions,13
  diode,3-13
  infrared,3-13
  IR,3-13
  planar diffused,26
  side emission,18,32,58,184
Encoder disc mounting,147
Encoder,99,132,135,138-149, 224,228,233,238,250
Encpasulation (see Package)
Epitaxial process,7-11
  buried layer,103
  deposition,89
  doping level,10
  gaseous phase,101
  GaAlAs,10-11
  GaAs,7-10
  liquid phase,7-11
  thickness,10
Eye switch,259

Fall time (see Switching speed)
Fiber optics,193-196
  advantages,195
  disadvantages,196
  graded core,193
  numerical aperture,194
  step index,193-194
  wavelength,193
Flash,25
Flow meter,239
Fluid level detector,233-234

Gain,90
Gallium:
  aluminum arsenide,5

Index

arsenide,5
arsenide phosphide,5
phosphide,5
Garage door opener,249
Gas pump,240
Gaseous phase,101
Glaucoma test,257-259
Golf game,254
Guard ring,103,106

Head up display,248
Helicopter landing lights,246
Homing sensor,246
Hot tub remote control,249
Housing materials,149-150
Hypersonic missle,246

IRED (see Emitters)
IRLED (see Emitters)
IV drop monitor,257-258
Ice detector,245
Impurity:
  concentrations,7
  IC,101-107
  IRED,7
  photodiode,69
  phototransistor,81
  profile,8,9,11,70,82
Incidence response,77,92
Infrared film,254
Infrared landing system,246
Integrated circuit,99-113
  capacitor fabrication,109
  oxide overcoating,109
  resistor fabrication,105-108
  transistor fabrication,108
Inverse square law,40-41,177,197,206
Ion implantation,107

JEDEC,175
Joystick,249-250
Junction heating,216
Junction:
  diffused,8
  NPN phototransistor,85
  PN,3-5,69
  side radiation,18,58
  temperature,215

Keyboard,226-227

LED (see Emitters)
Laser disk,257
Laser,193,198,201
Leakage,80
Length checker,233,237
Lens,23,32,97-98,133,156,197,226
Light pen,231
Light pipe,161,168,180,184,187,231
Linear operation,215
Liquid level detector,233-234
Liquid phase,7-11
Load leveler,242-243

MOS,109
Margin control,224
Mark sense detectors,154,160, 231
Memory disk,224-225
Metalization process,103,106-107
Metalization,28
Meter reader,249
Morse code,198
Mouse,228

Noise:
  diode,77
  flicker,77
  NEP,77,80
  photo IC,99
  shot,77
  thermal,78

Ohm's law,213
On-axis intensity,58
Open air communication,197-201,228
Optical isolator,167-169
  AC input,178
  applications,232,240
  controlling agencies,167-169
  degradation,173-174

fabrication,184-189
high isolation voltage,
  179-180
isolation voltage,168
logic output,182
triac driver,121-123,178,
  255
waveforms,169-171
Optical measurement,237
Optical tracking control,251
Optocoupler (see Optical
  isolator)

PDIP (see Optical isolator)
PDIP counter,239
PIN diode,71,199-201,218
Package:
  aperture,133
  comparison,24,31
  construction,19-21,24,30
  dimensions,32
  hermetic,18,39
  molding process,23
  optical efficiency,18
  optocoupler,172-176
  plastic,18,36-39
  ruggedized,39
  T-1 3/4,33
  TO-46,33
Paper edge,226
Paper path sensor,230-231
Paper presence sensor,224,
  230-231
Paraplegic,259
Pattern Plots,33-34,36,57
Phonon,4
Photodarlington,222
Photodiode,69-80
  amplifier circuit,74
  capacitance,71-72
  efficiency,75-77
  indicence response,77
  noise,77
  photoconductive mode,72
  photovoltaic operation,
    69,72
Photoelectric effect,69
  conduction band,69
  depletion region,70
  valence band,69
Photon,4,53,70,75,217

Pill counting,257
Pilot eye switch,248
Plethismograph,257
Polycarbonate,149-150
Polysulfone,149-150
Pool ball sensor,254
Power dissipation:
  envelope,47-48
  maximum IRED,46
Power output:
  vs. angular displacement,
    33,34-36,57
  degradation,35,215
  vs. forward current,42
  measurement technique,32,
    53-62
  parameters,57
  pulsed,202-203
  vs. temperature,42
  vs. time,54,62-65
  units,53-62
Printer,224
Pulse counter,257
Pulse operation,202-210
Pulse receiver,208

Quadrature,144-146

Recombination,14-15,69,217
Reel size detector,229
Reflective optical switch,
    154-164
  cross talk,162
  customization,161
  diffused reflectance,155
  dust cover,162
  focused assemblies,156,161
  liquid level sensing,157-
    159,233
  polished surface,155
  sufrace variations,158
  unfocused assemblies,163
Refractive index:
  matching compound,14
  N1/N2 interface (see
    Critical angle)
  silicone gel,17,22,184-189
Reliability,63-65,203
Remote control door locks,
    244

Index

Remote control, 197, 201, 244
Reverse breakdown, 86
Ring detector, 260
Rise time (see Switching speed)
Robotic encoder, 233, 238

Safety shield, 233-234
Schmitt trigger, 112, 181
Security system, 233, 235
Sewing machine, 240
Shutter requirements, 127-132
Silicone gel, 17, 22, 184-189
Slack length detector, 229
Slot machines, 253
Slotted switch (see Transmissive switch)
Smoke detector, 240, 252
Snapback, 87
Snell's law, 149, 159, 193
Snubber network, 122
Solar cell, 72
Solid state relay, 240
Spectral response, 92
Sunlight, 75
Surface detectors, 154-159
Swimming pool control, 249
Switching power supplies, 246
Switching speed:
  couplers, 169, 173
  comparison, 99, 110, 133, 222
  encoder, 132
  IRED, 43, 74
  vs. load resistance, 96, 220
  mechanical, 129-132
  photo IC, 109
  photodiode, 73-74, 217
  photodarlington, 83, 97
  phototransistor, 94

TTL output, 182
TV remote control, 249
Tape reader, 229
Telephone interconnect, 260
Television, 249, 255-256
Temperature:
  vs. collector current, 94
  coupled output, 94, 174-175, 213
  and current, 28
  vs. dark current, 79, 91
  vs. degradation, 63
  glass transition, 38
  junction, 45, 215
  operating, 38, 220
  vs. output, 42, 54
  storage, 37
  shock, 38
  vs. wavelength, 44
Textile mills, 240
Thermal impedance, 45-52
  envelope, 47
  measurement, 45, 49
  rating, 46
Three dimensionsl television, 255-256
Tone arm return, 252
Torpedo sensor, 246
Touch screen, 226-227
Track ball, 249-250
Track zero, 224
Tractor feed, 226
Traffic light control, 240
Transmissive switch, 127-153
  aperture, 135
  blocking media, 127, 132
  custom, 152
  flag, 127-132
  housing materials, 149-150
  on/off conditions, 128
  output waveform, 130-131
  potentiometer, 136
  sensitivity matching, 137
  saturation, 129
  speed, 132
Triac driver IC, 121-123, 178, 255

Vending machines, 253
Voltage:
  isolation, 168
  offset, 90
  reverse, 71
  saturation, 90
  zero crossing, 178

Waiter order placement, 249
Wavelength, 10
  fiber optic, 193

vs. photon efficiency, 53
vs. sensor response, 72-73
variations, 43
visible, 5
Winchester drive, 224-225

Write protect, 225

Zero referencing:
  encoder, 147
  track zero, 193-196